T0257537

Water Quality Management and Pollution Control:
A Global Overview

Edited by **Bruce Horak**

New York

Published by Callisto Reference,
106 Park Avenue, Suite 200,
New York, NY 10016, USA
www.callistoreference.com

Water Quality Management and Pollution Control:
A Global Overview
Edited by Bruce Horak

© 2015 Callisto Reference

International Standard Book Number: 978-1-63239-611-2 (Hardback)

Contents

Preface

Over the recent decade, advancements and applications have progressed exponentially. This has led to the increased interest in this field and projects are being conducted to enhance knowledge. The main objective of this book is to present some of the critical challenges and provide insights into possible solutions. This book will answer the varied questions that arise in the field and also provide an increased scope for furthering studies.

Water is a prerequisite for any form of life. The level of surface water quality protection is uneven around the world due to the relative efficiency of ecological instruction and the scale to which science influences the regulatory procedure. In the US, the Total Maximum Daily Load (TMDL) has been a triumphant policy for water quality treatment process. The TMDL offers a balanced structure for estimating the assimilative capability of the receiving water body for certain contaminants, applying factors of security and incorporating suitable levels of water quality criteria violation - provided the local stakeholders have a say in the decision making process. This book is a compilation of all such researches which will help our readers in understanding the topic better.

I hope that this book, with its visionary approach, will be a valuable addition and will promote interest among readers. Each of the authors has provided their extraordinary competence in their specific fields by providing different perspectives as they come from diverse nations and regions. I thank them for their contributions.

Editor

Endocrine Disruptors in Water Sources: Human Health Risks and EDs Removal from Water Through Nanofiltration

J. E. Cortés Muñoz, C. G. Calderón Mólgora,
A. Martín Domínguez, E. E. Espino de la O,
S. L. Gelover Santiago, C. L. Hernández Martínez and
G. E. Moeller Chávez

Additional information is available at the end of the chapter

1. Introduction

Water is fundamental for human health and well-being as well as for stimulating diverse socioeconomic activities. Paradoxically, these very activities have contributed to the alteration and deterioration of water supply sources from a microbiological, physical and chemical standpoint, causing sanitary risks for the population. For example: since the end of the 19th century, the role of drinking water in exposing populations to pathogens, and improvements in its quality in order to prevent diarrheic illnesses, has been widely analyzed, debated and documented [1,2]; in the 20th century, epidemiological evidence was found of cutaneous lesions [3]and various types of cancer related to hydroarsenicism [4], as well as dental and skeletal fluorosis related to fluoride in drinking water [5].

In recent decades the problem of these possible public health risks from so-called emergent contaminants (ECs) has been factored into the problem that includes a wide range of compounds whose environmental presence and impact have been proven with the advent of new sensitive and reliable quantitative analytical tools [6]: ECs are bioactive substances synthesized and used for the household, agriculture, livestock, industry, personal care products and hygiene (PCPs), and human and veterinary medicine, including byproducts of production and degradation [7].However, beyond the concentrations and environmental persistence of ECs, their relevance lies in the fact that they continue to be released into the environment through various ways which confers ubiquity.

Chemical endocrine disruptors (EDs), which are xenobiotics (compounds exogenous or foreign to living organisms) with potential to alter hormonal regulation and normal endocrine system function, consequently affecting an intact organism, its progeny or subpopulations, are among the wide range of ECs [8]. The evidence of adverse effects on aquatic organisms at relevant environmental concentrations [9,10]is well documented as well as *in vitro, in vivo* and epidemiological studies that associate human exposure to these compounds with: obesity, metabolic syndrome, type II diabetes mellitus [11],estrogenic, androgenic and antiandrogenic activity or combinations thereof [12], reproductive, nervous and immune systems effects, as well as some cancers and developmental effects [13].

The presence of EDs in bodies of water is due mainly to the discharge of wastewater that impacts the quality of surface water and groundwater with compounds that are not entirely removed by conventional treatment processes [14,15], which is particularly relevant in areas like the Mezquital Valley (state of Hidalgo, Mexico).The aquifer that supplies the population is recharged with the residual waters used in agricultural irrigation. Another way is through the indirect reuse of treated wastewater for potable water source augmentation. These practices could explain why some pharmaceuticals (Phs) and personal care products (PCPs) have been detected in waters treated for human consumption [16].

The concern regarding human exposure to EDs, in this case through water consumption, is based on five points: 1) evidence of adverse effects on fish and aquatic ecosystems at relevant environmental concentrations [9,10]; 2) documented clinical cases of cancers related to hormones in industrialized nations [8], as well as prevalence of reproductive disorders in adolescents and young men in Europe [17]; 3) *in vivo* studies that show endocrine disruption through exposure to certain ambient chemicals; 4) various chemical compounds classified as EDs or with potential to act as such, have been found in surface water and groundwater [18]and, 5) evidence that suggests that conventional water treatment systems are inefficient in removal of these types of contaminants [16].

The European Union, Germany, England, USA, Australia, Canada and Japan have all installed multi-stage treatment systems that effectively reduce the concentration of EDs in drinking water. The debate has begun over the need for research and regulation, analytical methods, water sources and treated water monitoring, public health and environmental risks, water treatment processes, transformation, transport and fate in the environment of EDs.

In Mexico there are few studies related to the occurrence of EDs in water, as well as few studies that document the efficacy and efficiency of water treatment processes in the removal of ECs and EDs. Considering that the Mezquital Valley is a prime example of an aquifer affected by the reuse of wastewater, and that the occurrence of ECs has been documented in supply wells in the area, the following objectives are proposed:

a.   Analyze and synthesize information regarding the presence of EDs in supply sources and treated waters for potable use, sanitary, environmental and regulatory relevance, treatment processes for removal from water.

b. Analyze the problems related to ED exposure, specifically arsenic, bisphenol A, alkyl-phenols and their ethoxylates through the use and consumption of water in the Mezquital Valley.

c. Analyze the technical feasibility of nanofiltration (NF) process to remove mineral and organic compounds from groundwater in the area of the study.

d. Identify investigation needs regarding the occurrence concentration, persistence, transformation and destination, action mechanisms and risk assessment of EDs in water for human consumption and treatment processes for potable use

2. Endocrine disruptors in drinking water: Public health relevance

By May 16, 2012, the Chemical Abstracts Service [19], had registered over 66.67 million organic and inorganic substances. More than 100,000 man-made chemicals are available on the market including approximately 1,500 new molecules released yearly [20,21] for manufacturing products whose primary use is for human well-being and socioeconomic development. Since the 1990s, EDs have been one of the most controversial issues, attracting the attention of the scientific community, international agencies and organizations, governments and the general public.

The U.S. Environmental Protection Agency defined an endocrine disruptor as "an exogenous agent that interferes with the production, release, transport, metabolism, binding, action, or elimination of natural hormones in the body responsible for the maintenance of homeostasis, reproduction, development, and/ or behavior" [22]; the European Commission defines it as "an exogenous substance or mixture that alters function(s) of the endocrine system and consequently causes adverse health effects in an intact organism, or its progeny, or (sub)populations" [8]. From both definitions it is clear that EDs are compounds that alter hormonal regulation or homeostasis that can cause undesirable adverse effects on health as a result of exposure to a compound whose mechanism or action is endocrine disruption.

2.1. Origin and occurrence of EDs in drinking water

The nature and origin of EDs is diverse and includes groups of compounds such as: active ingredients in medicines with collateral hormonal effect, pesticides and adjuvants for their application, products to increase growth and weight gain in livestock, personal care and hygiene products, flame retardants, chemicals for use in the plastic industry and other frequently used industrial chemicals, natural and synthetic hormones, as well as products for manufacturing consumer goods and degradation by products [20,22-26].

Their impact on public health and wildlife is due to their bioactivity and ubiquity in the environment, as they are introduced unconsciously and permanently in the various environmental compartments. They can be introduced as pure substances or complex mixtures through diverse ways, especially via the flow of treated or untreated wastewaters. These compounds are not totally removed or inactivated by conventional water treatment systems or by natural processes of self-purification of the receiving bodies (water or ground), frequently reaching groundwater [26-28].

For example, in the Mezquital Valley, the aquifer that supplies the population is recharged mainly with wastewaters from Mexico City from agricultural irrigation; however, while the contamination with EDs, Phs, and other organic compounds originate from these wastewaters, they also originate from the disposal of PCPs and expired or unused medications by households, municipal and hospital wastewater, leachates from landfills and local uncontrolled garbage dumps (Figure 1, adapted from [29]), which is consistent with information published by various authors for other aquifers [29-31].

In this case, as in scenarios of direct or indirect reuse of treated wastewater as water supply source, the main concern is related to the pathogens as well as nitrates, Phs, PCPs and disinfection byproducts with potential to disrupt the endocrine system and affect human and environmental health [29].

In comparison to other chemical compounds, there is little information on the transformation and fate of EDs especially regarding biotransformation, hydrolysis and photo transformation of Phs and PCPs. Their low volatility suggests that their distribution in the environment will occur mainly via aqueous transportation and dispersion through the food chain. The polar and non-volatile nature of Phs impedes their release from water [31], without geographical or climatic borders for these synthetic substances that have been found in areas that are considered to have low pollution levels [32].

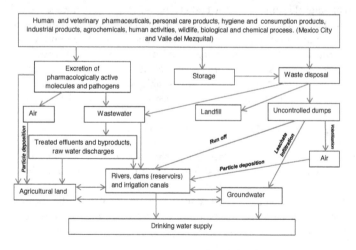

Figure 1. Conceptual model of introduction and transportation of ECs and EDs in the Mezquital Valley.

ECs, EDs and PCPs have been frequently detected in effluents and surface waters that could be present in drinking water. Ultra-trace concentrations (ng/L) of prescription and non-prescription Phs and their metabolites have been reported in samples of drinking water across the United States. These include caffeine, analgesics, anti-inflammatories (naproxen), anti-convulsants (carbamazepine, phenytoin, primidone) and anxiolytics (meprobamate), xray contrast medium (iopromide), lipid regulators (gemfibrozil), antibiotics or their metabolites, nicotine metabolites, hypotensives (atenolol), synthetic musk, a polycyclic aromatic hydro-carbon compound, a plant sterol, plastic components, an insecticide, surfactants (bisphenol-A, alkylphenols) and degradation products, a fixative used in perfumes and soaps, a flame retardant and a pesticide [16,28,33-36].

In general throughout the water cycle there is a reduction in EDs through absorption, dilution, and biodegradation [24], and yet there are still questions about their fate in the environment. Laboratory studies have shown that bioactivity is reduced over a period of hours to days due to degradation and sorption yet field studies indicate that estrogens are sufficiently mobile and persistent to impact the surface and ground waters [37], while in the ground and sediments where they can persist for long lengths of time (the half-life of clofibric acid, for example, is estimated at 21 years), they reach levels in the g/kg range [23].

Given the ubiquitous nature of EDs, all humans are exposed through different tracks: inhala-tion, ingestion and dermal contact. Contributing to this is their low biodegradability, air and water transportability, bio accumulation in the trophic chain and transgenerational exposure; fetuses are especially vulnerable because pregnant women accumulate EDs in adipose tissue [38].As with other environmental chemicals, their effects depend on the concentration and nature of the chemical as well as its route, frequency and intensity with which exposure occurs and in this case the phase of life at which the exposure occurs.

Due to the trace and ultra-trace concentrations in which EDs are found in drinking water, people are commonly exposed in higher quantities through medications and other sources and routes: diet, inhalation of airborne chemical substances and dermal absorption (topical medication or personal care products), which suggests the contribution by drinking water to the overall exposure and its relative importance in assessment of sanitary risks associated with these types of contaminants [39].

Likewise from a risk management viewpoint, it is important to note that the variety and chemical structure of EDs complicates their identification and quantification in water as well as the characterization of the sanitary risks associated with chronic exposure to low or environmentally relevant doses In addition little is known about the occurrence, toxic-ity and potential endocrine activity of the products of degradation that may result from the processes of bio and physiochemical transformation that alter the chemical structure of EDs rather than eliminate them.

2.2. Human health effects

2.2.1. Endocrine system

Hormones are produced by the glands that comprise the endocrine system, which is the key to communication within the human organism and control method between the nervous system and the various functions of the human body such as reproduction, immunity, energy control, metabolism, conduct, growth and development. For example:

- The Thymus gland is the source of immunologic regulatory hormones [35].

- The Hypothalamus gland releases hormones such as oxytocin; it's important in the control of reproductive endocrine processes and stimulate pituitary activity [35,40,41].

- The Pituitary gland releases steroid hormones, such as corticosteroids, androgens, and estrogens; growth hormone, oxytocin. Feedback control signals the endocrine organ (the adrenal gland or gonads), to cease production or release of the endogenous steroid or to stimulate the release of opposing hormones. This homeostatic control in response to endogenous hormones is critical for maintaining proper hormone concentrations [35,40].

- The Thyroid gland releases thyroid hormones (calcitocin and thyroxine), which are receptor nuclear of steroids, regulate metabolism, growth, development, behavior and puberty [35,40,41].

- The Adrenal glands, release corticosteroid hormones, cathecolamines to regulate metabolism and behavior [40,41].

- The Pancreas produces insulin and glucagons to regulate blood sugar levels [40].

- The Ovaries and testicles produce sex steroids such as estrogen, progesterone, testosterone (androgens and estrogens) [35,40].

This means that all the physiological systems sensitive to hormones are vulnerable to EDs, including the brain and hypothalamic-neuroendocrine systems, cardiovascular system, mammary gland, adipose tissue, ovary and uterus in females, and testes and prostate in males [25,40].

2.2.2. Endocrine disruption mechanisms

Hormones circulate in the blood stream to modulate cellular and organ function through the union with complex molecular receptors and mechanisms:

- They mimic natural hormones; if exposed to relatively high doses, they join receptors within the cell and block or interfere with the ways through which hormones and receptors are synthesized or controlled [42].

- Binding and activating estrogenic and androgenic receptors N40. There are a number of estrogenic receptors in gonads, liver, brain, and sex organs; union without activation of the receptor would act like anti-estrogenics or anti-androgenics [43-45].

- Binding without activation of the receptor would act like anti-estrogenics or anti-androgenics [43-45].

- Modifying hormonal mechanisms [46], or of the number of hormonal receptors in the cell, or of the production of natural hormones, for example in the thyroid, immune or nervous systems [47].

• Selectively inhibiting DNA transcription; for example, arsenic produces a disruption in the transcription of DNA induced by the glucocorticoids mediated by receptors [47].

The mechanisms described in recent literature also include: a) the alteration of the transcriptional activity of nucleus receptors by modulating co-regulators through mediated degradation of the proteasome as well as by inhibiting histone deacetilase activity and stimulating mitogenic quinase protein activity; b) regulation of the methylation of DNA and c) modulation of lipid metabolism and adipogenesis, which possibly contribute to the current epidemic of obesity [47-50].

2.2.3. Health effects

EDs are structurally similar to many natural hormones; some have lipophilic properties [40], act in extremely low concentrations and therefore can have effects on organisms with low dose exposures [50,51]. As a result, environmental presence in trace and ultra-trace amounts, particularly in water, may be insufficient to cause cellular death or act upon genetic material, yet could result in a source of human exposure and carry sanitary risks for more susceptible segments of the population.

The target hormones for EDs and the effects differ from one compound to another (as shown in Table 1), as well as among species and intra species; for example, there are reports that the median Bisphenol A (BPA) level in human blood and tissues, including in human fetal blood, is higher than the level that causes adverse effects in mice [52].

The time of exposure to EDs in the organism in development is decisive in determining its character and future potential and, even when critical exposure takes place during embryonic development, the clinical manifestations may not be present until adulthood [8,43,45,46]. One compound may act on different target hormones [40] and cause different alterations. A wide and current review of the health risks is found in [22,53]. Some examples of compounds intensively used in Mexico, and thier potential effects in in humans, are presented in Table 1, as well as their normal application or use and the exposure source. It is worth mentioning that the exposure is involuntary.

2.3. NF as an alternative to remove ECs from water

Between January 2001 and July 2004 the European Union conducted Project Poseidon [64]; among its objectives it proposed conducting integral studies to evaluate and improve the removal of pharmaceuticals and personal care products from residential residual waters using conventional and advanced treatments as well as with potable water. One of the conclusions of the study is that reverse osmosis, nanofiltration and ultrafiltration-powdered activated carbon are powerful processes for the removal of pharmaceuticals and personal care products, among which are found ECs and EDs (suspicious and recognized).

In spite of the conclusions from the POSEIDON Project, the question about the efficacy of NF membranes in removal of emergent contaminants persists. Many studies have been made of this topic [65-73]. The spectrum of tests covers membranes with a molecular weight cut off (MWCO) of 200, 400 and 600 Da, as well as organic compounds with different molecular weights, sizes and physiochemical characteristics.

Example	Use	Involuntary exposure source	Example of adverse effect in humans
Bisphenol A	Plasticizers	Plastic items, packaging foods, beverages, water	Estrogenic [52-55], thyroid hormone, progesterone[40], androgenic. Female breast structural anomalies, premature thelarche, cancer; pubertal timing variations, ovulatory disorders, sexual behavior, prostate cancer [56,57]. Disrupted hypothalamic estrogenic receptor distribution, altered nitric oxide syntheses signaling [57]. Affectation of human immune function [58], and disrupted behavior in children associated with early life BPA exposure, especially in girls [59].
Vinclozolin	Fungicides	Foods (fruit, vegetables, cereals), water	Estrogenic, anti-androgenic. Promotes transgenerational adult onset disease such as male infertility, kidney and prostate diseases, immune abnormalities and tumor development [60].
Alkylphenols	Detergents, emulsifiers, agrochemicals	Household items, water, foods (fish)	Estrogenic [40], androgenic. Environmental and health issues continue to cast uncertainty over the human risks of alkylphenols and alkylphenols ethoxylates.
Phthalates	Plasticizers	Plastic items, cosmetics, personal products care, beverages, water	Estrogenic, androgenic, thyroid hormone [40]. Abnormalities such as hypospadias, cryptorchidism, reduced anogenital distance. Oligospermia, germ cell cancer [56]. Neurodevelopment and metabolic endpoints are of concern, since studies of prenatal exposure have found associations with phthalate exposure and lowered IQs, and exposure has been implied as a risk factor for obesity, insulin resistance and diabetes by others [47,50,55].
Diethylstilbestrol	Contraception, hormone replace therapy	Pharmaceuticals. They have been detected in water sources, wastewater and treated effluents	Estrogenic. Abnormalities such as hypospadias, cryptorchidism, reduced anogenital distance. Female structural anomalies, breast cancer, structural anomalies, premature thelarche. Prostate cancer [56,61].
Ibuprofen, diclofenac, acetaminophen	Anti-inflammatory, analgesics		Candidates may be identified on the basis of simple assumptions regarding their use and activity: a) non estrogenic steroids may react with environmental endocrine receptors or metabolize on their way to the environment and thus form endocrine disruptors; b) many high-volume drugs released to the environment have not yet been tested for their endocrine properties, and some of these are known to interact with the human endocrine system [62].
Bezafibrate, clofibrate, gemfibrozil, fenofibrate	Lipidic regulator		
Atenolol, metoprolol, propanolol	B-blockers, antihypertensives		
Anabolic steroids; trenbolone acetate, melengestrol acetate	Fast growth of meat producing animals	Animal food, soil, wastewater	Androgenic. A placebo-controlled prospective study demonstrated adverse and activating mood and behavioral effects of anabolic steroids [63].

Table 1. Example of possible sources of exposure to EDs and target hormone system.

NF efficiently removes dissolved solids, organic carbon, inorganic ions and organic micro contaminants, both regulated and unregulated, and therefore its capacity is similar to that of reverse osmosis (RO); however, due to higher pressure requirements the latter has higher investment and operational costs. Furthermore the partial discrimination of calcium and bicarbonates of the NF can be advantageous since drinking water distributed through municipal networks should be saturated with calcium carbonate to avoid corrosion [74].

The mechanisms through which RO and NF reject organic compounds are mainly: a) exclusion by size or stericity; b) exclusion by repulsion of charges or ionic exclusion also known as Donnan Effect and c) the physiochemical interaction of the solvent, the solute and the membrane [75].

Size exclusion is associated with the molecular weight cut off of a membrane but this is relative to a certain type of molecule, therefore other molecules with similar molecular weight yet different physiochemical properties may not be rejected at the same rate. In the case of neutral compounds, other geometric descriptors of the molecule are far more useful such as molecular width [76]or hydrodynamic radius, especially for molecules whose radius approaches that of the membrane pores [77].

Exclusion by charge repulsion is a phenomenon in highly typified NF membranes; tests conducted with NF membranes and ultra-low pressure reverse osmosis to retain or reject 9 organic compounds (5 negative ionic charge and 4 neutral) showed that the electrically negative charged compounds were rejected with efficiencies above 91% even with molecular weights lower than the rated molecular weight cut off of the membranes, while the neutral compounds were rejected at much lower rates at levels as low as 12% for 2-naphtol [78].

The interaction between solvent, solute and membrane can lead to misleading results if enough time is not allotted for the process to normalize, since various organic contaminants tend to be adsorbed on the membrane surfaces. As a result, initially they are retained, but once the membrane surface is saturated, the true rejection of the compound can be observed [65,77-81].

NF membranes (MWCO of 200, 300 and 400 Da) and RO membranes (MWCO not reported) have a 90% reject rate of ionic and non-ionic hydrophilic organic compounds such as naproxen, diclofenac, ibuprophen, mecoprop, ketoprophen, gemfibrozil, and primidone and higher for NF membranes with MWCO of 200 Da as well as the RO membranes. Even the NF membrane with MWCO of 300 Da had reject rates of over 90% [68].

Tests conducted with a membrane with a MWCO of 300 Da (Filmtec NF200) to evaluate the removal of hormones and antibiotics with microfiltration showed that in a matrix of drinking water with a mixture of hormones and medications (i.e. hormones with sulfonamides or tetracycline) it is possible to obtain rejects of nearly all of the substances with the exception of testosterone whose reject rate was 95%. The pharmaceuticals tested were tetracyclines, (chlorotetracycline, sulfachloropiridazine, sulfamerazine, sulfametoxazole, sulfametazole) and hormones (estrogen, progesterone, testosterone and 17α-ethynilestradiol). There is also evidence that the removal efficiency of hormones and sulfonamides was lower when deionized water was used and a pure solute [72].

As shown in the previous examples, the results obtained in the laboratory tests on efficiency of NF membranes in the removal of ECs have been very favorable even though it has been shown that compounds with a lower molecular weight, less branched and without ionic charge, were only partially retained.

In pilot applications or full scale, the performance of NF in the control of organic micro contaminants is less documented; even so, there are two pilot systems documented [67, 69] and one municipal plant [73]. The main results of these two applications are described below.

A comparative study between ultra-low pressure reverse osmosis membranes (ULPRO) and NF showed that both types of membrane can reach efficient levels of removal of organic compounds such as pharmaceuticals, pesticides, flame retardants, plasticizers, and nitrogen, similar to that of conventional reverse osmosis, producing water with equal quality as required for indirect potable reuse of treated water. The operational conditions and results show that both types of membrane are viable for reuse projects of treated residual water in which a high quality permeate is required [67].

A NF pilot plant with a capacity of 56.4 L/min was fed with water from an advanced treatment facility for residual water. After the biological treatment, the water was micro filtered and disinfected. The pilot plant operated continuously for 1200 hours (50 days). Ten emergent contaminants were found in the source water (3 plasticizers, [TCEP, TCPP, and TDCPP], acetyl salicylic acid, naproxen, ketoprophen, diclofenac, gemfibrozil, primidone, and carbamaze-pine) and in all cases removal rates of 75-100% were observed, except during the first 24 hours of operation when efficiencies were lower. However, as the test progressed, an increase in removal of contaminants was observed to the degree that the organic compounds with ionic charge and the neutral ones with greater molecular weight were rejected at a rate of 90%/.

A research study was held at a municipal drinking water facility located in the north of Spain. The plant has 3 parallel trains of membrane processes, two of which are reverse osmosis (486 m³/h each one) and one NF (360 m³/h).The facility is fed from groundwater wells, directly influenced by infiltration from Besós River. During the study, five meas-urement campaigns, aimed at 12 pharmaceutical compounds, were conducted. The re-sults obtained indicate that NF was capable of removing hydrochloratizide, ketoprofen, diclofenac, sotalol, sulfamethoxazole, metopropolol, propifenazone and carbamazepine with efficiency levels above 90%.

Compounds with negative charges like gemfibrozil and mefenamic acid had a low removal rate both in the NF (50% and 30% average), as well in RO (50% and 70% average). Meanwhile, acetaminophen, which has a molecular weight cut off of 151.16 Da., had a removal rate of 44.8% through NF and 73% through reverse osmosis. The authors conclude that the NF and reverse osmosis membranes applied in full-scale are very efficient in removal of almost all pharma-ceutical residuals found in water.

Due to advantages NF has over RO, such as lower energy consumption and selective ion discrimination, as well as the capability to remove almost all pharmaceutical residuals found in water it was decided to test a NF process for water treatment at Mezquital Valley.

3. Endocrine disruptors in groundwater sources of Mezquital Valley

3.1. Description of the study site

The Mezquital Valley is located in the central plateau of Mexico, 80 km. from Mexico City in the South central part of the state of Hidalgo. In the municipalities located within the limits of the aquifer of Mezquital Valley (Figure 2),there is an estimated population of 377,951 inhabitants, which represents 14.8% of the total population of the state [82].The principal economic activity in the region is agriculture, which is distributed in two irrigation districts (03 Tula and 100 Alfajayucan). In the area there are also a petroleum refinery plant, a thermoelectric plant, textile factories and a cement plant, considerable commercial activity, craft production, and the tourist industry is based on spas and hiking in the mountains [83].

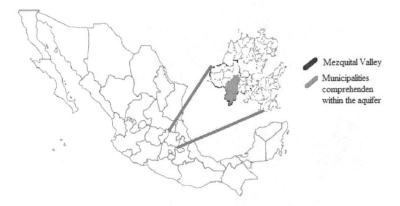

Mezquital Valley

Municipalities
comprehenden
within the aquifer

Figure 2. Location of the Mezquital Valley aquifer in Hidalgo State.

The climate is semi-arid, with a high incidence of sunlight and warm temperatures [83]. [83].Average annual precipitation is 450 mm, while evaporation is 2,100 mm; rain is limited to the months of May through October [84].

Within the Mezquital Valley are the irrigation districts 03 Tula and 100 Alfajayucan, representing one of the largest irrigation schemes of residual water in the world. It receives between 50 and 60 m^3/s of water (80% residual and 20% pluvial) from Mexico City through the Gran Canal del Desagüe, the Interceptor Poniente, and the Emisor Central (Figure 3)[85,86].

Around 75% of this water [87]is used without formal treatment for irrigation of approximately 85,000 hectares of saline soils that lead to high levels of irrigation (1.5 to 2.2 m) [86].Approximately 10,000 hectares receive crude residual waters directly, 35,000 hectares receive mixed water (80% residual and 20% from reservoirs and pluvial sources) and 25,000 hectares receive self-purified residual water from the Requena, Endho, Rojo Gomez and Vicente Aguirre reservoirs [88].

The Emisor Central and Emisor Poniente join the Tula River to feed the Endho reservoir (approximately 200 million m^3) El Gran Canal empties into the Salado river and reaches the

Tula river beyond the dam [88] (Figure 3). 81% of the main canals and 52% of the lateral canals are not improved or coated, permitting the infiltration and artificial recharge of the aquifer. The water outlets from the aquifer are through springs that lead to the Tula river, extraction through wells, and the rest discharges toward the north and northwest, both superficially (drained by the Salado river) as well as subterranean [85]. The estimated outflow from the springs, which that spread out within the municipalities in the region, range from 100 to 600 L/s, and supply the population [86].

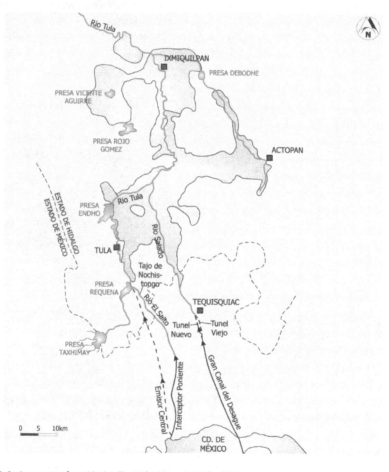

Figure 3. Sanitary sewers from Mexico City to the Mezquital Valley [85].

In the wastewaters, wells and springs in the Valle, sampling has shown traces of volatile and non-volatile organic compounds, phenolic compounds, and some PAHs and PCB's in concentrations in the order of picograms/L. Likewise, the presence of benzylbuthylphthalate, diethylhexyl phthalate, nonylphenol, salicylic acid, carbamazepine, ibuprophen and naproxen, indicate contamination from Phs and PCPs that include recognize and suspicious EDs compounds [87,89,90].

3.2. Sampling sites

For the first monitoring campaign, 19 supply sources were selected (Figure 4) located within the polygon that outlines the aquifer of the Mezquital Valley, and distributed throughout the irrigation zone with untreated wastewater, mixed waters (wastewater and fluvial reservoir water) and at a site to the north of the region to be used as a control site. The 11 sampling sites of wastewater include the water that flows into the Valle via the Emisor Central, residual waters from irrigation canals and one point in the Xotho canal that carries agricultural runoff waters as a point of comparison (Figure 5).

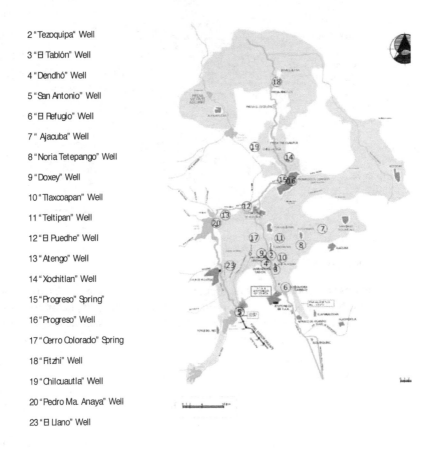

2 "Tezoquipa" Well

3 "El Tablón" Well

4 "Dendhó" Well

5 "San Antonio" Well

6 "El Refugio" Well

7 " Ajacuba" Well

8 "Noria Tetepango" Well

9 "Doxey" Well

10 "Tlaxcoapan" Well

11 "Teltipan" Well

12 "El Puedhe" Well

13 "Atengo" Well

14 "Xochitlan" Well

15 "Progreso" Spring"

16 "Progreso" Well

17 "Cerro Colorado" Spring

18 "Fitzhi" Well

19 "Chilcuautla" Well

20 "Pedro Ma. Anaya" Well

23 "El Llano" Well

Figure 4. Supply sources.

W2 Emisor Central. Pipe/canal

W3 El Salto. River

W4 Salado. River

W5 Tlamaco-Juandho. Canal

W6 Salto-Tlamaco. Canal

W7 Salto-Tlamaco 2. Canal

W8 Requena 1. Canal

W9 Dendho. Canal

W10 Requena 2. Canal

W12 Requena 3. Canal

W13 Xotho. Canal

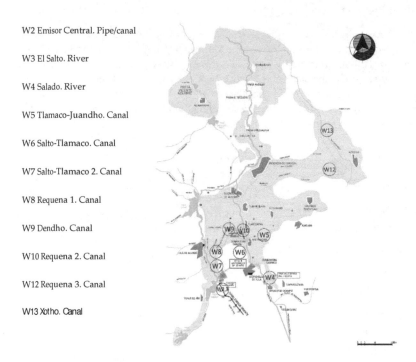

Figure 5. Wastewater sample sites.

A second sampling was later conducted on eight supply sources to test for unregulated organic compounds, phenolic compounds, estrogenic hormones, and a screening for pharmaceuticals, personal care products and progestogenic hormones; a second more specific and selective sampling was conducted of the supply sources in accordance with the following criteria:

- Concentrations of arsenic, fluorides, mercury or lead above maximum allowed limits as established by the Mexican normative for potable waters.

- Abundance (qualitative) of unregulated organic compounds (volatile, semi-volatile and persistent)

The sources selected were: 002 (Tezoquipa), 006 (El Refugio), 013 (Atengo), 014 (Xochitlan), 015 (Progreso spring), 017 (Cerro Colorado spring), 018 (Fitzhi) y 019 (Chilcuautla)

3.3. Sampling

The procedure for gathering water samples to analyze compounds at the trace and ultra-trace levels requires special care in order to not contaminate the samples by physical contact, perspiration or exhalation. For this purpose a cap, powder free gloves and face masks were used while handling, filling, closing, labeling and packaging the containers for storage and transport.

Two people collected the samples; one was designated "dirty hands" and the other "clean hands". All operations requiring contact with the bottle and transferring the sample from the collection vessel to the bottle were conducted by "clean hands," and "dirty hands" was responsible for all activities that did not involve direct contact with the sample [91].

3.4. Analytical methods

The samples were analyzed in laboratories accredited by the Entidad Mexicana de Acreditamiento, and the analytical methods utilized are based on the Mexican Standards or on international methods or those applied in other countries and were implemented, standardized and or validated in the selected test labs for sample analysis. For the analysis of volatile compounds, the samples were gathered in vials of 40 ml, and were kept at a temperature of 4°C during transport to the lab and until their analysis.

The samples were submitted for analysis through a methodology of USEPA Method 5030B [92],using a gas Chromatograph Varian 3800 coupled to a mass spectrometer Saturn 2200 equipped with a capillary column Sample Concentrator Teckmar/Dohrmann Model 3100 and auto sampler Varian Archon for 40 ml. vials.

The analysis of semi-volatile compounds was conducted with US-EPA Method 8270D [93]. Extractions were made in basic, neutral and acidic medium. The fractions obtained were analyzed with gas chromatography-mass spectrometry in total ion monitoring mode. For the analysis of noniphenols, method ASTM-D 7065-06 was applied. Dichloromethane was used to extract the samples. The extracts were concentrated and dried with the use of anhydrous sodium sulphate.

The detection of PPCPs was conducted by AXYS Laboratories in British Columbia Canada. The samples were analyzed following the AXYS method MLA-075: Analytical Procedures for the Analysis of Pharmaceutical and Personal Care Products in Solid, Aqueous and Tissue Samples by LC-MS/MS[91]. This method is suitable for the determination of a suite of pharmaceutical and personal care compounds in aqueous, solid and tissue samples. The analysis requires extraction at two different pH conditions: at pH 10 for analysis of fourteen analytes; and at pH 2.0 for the analysis of the other analytes. Prior to extraction and/or clean-up, samples are adjusted to the required pH and spiked with surrogates. A total of 119 analytes can be identified and quantified. Analysis of the sample extract is performed on a high performance liquid chromatography coupled to a triple quadrupole mass spectrometer. The LC/MS/MS is run in MRM (Multiple Reaction Monitoring) mode and quantification is performed by recording the peak areas of the applicable parent ion/daughter ion transitions. Some analytes are analyzed in the ESI positive mode and some are analyzed in the ESI negative mode.

3.5. Results

3.5.1. Fluoride and heavy metals: Suspected endocrine disrupters

The Mexican standard for drinking water quality [94], among other chemical parameters, limits the concentration of heavy metals, fluoride, organochlorated pesticides, total trihalomethanes and volatile organic compounds. In this framework, only four contaminants with potential to cause endocrine disruption exceeded the established maximum contaminant level: Arsenic (As), Fluorides (F), mercury (Hg) and lead (Pb) [95,96].

The greatest challenge came up in the wells in the south region of the Mezquital Valley (002 through 012 and 23), that are located in the entry zone of the residual waters. In the central and northern zone of the valley high concentrations of fluorides were found (Table 2), which is consistent with data reported in a previous study [85].

Parameter (Max allowable limit)	Supply Source (concentration µg/L or mg/L)
Arsenic (25 µg/L)	Wells: 003 (26), 004 (35.1) y006 (38.7)
Fluorides (1.5 mg/L)	Wells: 004 (1.72), 006 (1.52), 009 (1.62), 012 (1.59); Springs:015 (2.37), 017 (1.64); Wells: 016 (3.32), 018 (3.04), 019 (2.98) 023 (3.77)
Mercury (1 µg/L)	Wells:002 (2), 003(1.1), 006 (2.6), 007 (2.1) y 010 (2.6)
Lead (10 µg/L)	Wells:002 (28), 003(30) y 004 (39)

Table 2. Supply sources and parameters of the Official Norm for Water for Human Consumption and Use that are not met in the study zone.

Chronic exposure to As in drinking water has been strongly associated with increased risks of multiple cancers, heart disease, diabetes mellitus II and reproductive and developmental problems in humans [65].Recent studies suggest that increased human health risks [66],at levels as low as 5-10 ppb, could be mediated, at least partially, through the capacity to alter steroid receptors [97-99].There is evidence that high exposures to F⁻ are associated with decreased thyroid function [100,101], increased activity of the calcitonine and parathyroid, secondary hyperparathyroidism, tolerance to glucose and possible effects over time on reaching sexual maturity [96]. Lead (Pb) and mercury (Hg), can interfere with hormone neurotransmitters and other growth factors, and in both cases any exposure can be considered dangerous to the developing organism [57,102].

3.6. Unregulated organic compounds and estrogens

In the first monitoring campaign, 80 volatile and semi-volatile organic compounds (VSOCs) were identified at a qualitative level. The south zone of the valley showed greater diversity of organic contamination and in the north zone, where the irrigation water is mixed (wastewater and fluvial reservoir water), the water supply sources also showed organic contamination (Table 3 and Table 4), including naphthalene in well 019.

Water sources in the central zone (wells007, 012, 013 and015) contained the least diversity of organic compounds, nevertheless, no clear trend can be observed; therefore we assume that human activities within the Valley represent an important contribution to the organic loading (Table 3).

Supply source	Number of organic compound	Supply source	Number of organic compound
002 "Tezoquipa" Well	22	003 "El Tablón" Well	17
004 "Dendhó" Well	17	005 "San Antonio" Well	26
006 "El Refugio" Well	26	007 " Ajacuba" Well	18
008 "Noria Tetepango" Well	20	009 "Doxey" Well	17
010 "Tlaxcoapan" Well	17	011 "Teltipan" Well	22
012 "El Puedhe" Well	11	013 "Atengo" Well	17
014 "Xochitlan" Well	22	015 "Progreso" Spring	8
016 "Progreso"	24	017 "Cerro Colorado" Spring	22
018 "Fitzhi" Well	29	019 "Chilcuautla" Well	31
020 "Pedro Ma. Anaya" Well	20	023 "El Llano" Well	17

Table 3. Number of volatile, semi-volatile and fixed organic compounds detected in water supply sources

Sampling site	Number of organic compound	Sampling site	Number of organic compound
AR "002 Emisor Central" Pipe/ canal	56	AR 003 "El Salto" River	37
AR 004 "Salado" River	42	AR 005 "Tlamaco Juandhó" Canal	42
AR 006 "Salto-Tlamaco 1" Canal	56	AR 007 "Salto-Tlamaco 2" Canal	39
AR 008 "Principal Requena 1" Canal	38	AR 009 "Principal Dendhó" Canal	39
AR 010 "Principal Requena 2" Canal	35	AR 012"Principal Requena 3" Canal	32
AR 013 "Xotho" Canal	15		

Table 4. Number of volatile, semi-volatile and fixed organic compounds detected in wastewater

In the second monitoring campaign conducted during the rainy season all eight supply sources showed VSOCs (173 different ones). Qualitatively, the greatest contamination was observed mainly in the central zone; the source with the highest contamination was 015 with 69 VSOCs followed by 013 with 59 VSOCs; 017, 42; 006 and 014 with 41; 019, 38; 018, 36; and the source least contaminated was Tezoquipa (002) where 32 VSOCs were found.

From both groups of compounds, the following are notorious for their presence in residual waters and supply sources:

- Pesticides: 1,4-dichlorobencene, 1,2-dicloropropane, 3-chloro-2-methyl-1-propene, 1,1-dichloro-1-nitroethane, 1,2,3,4-tetrahydronaphtalene.

- Flavorings and essences: 3-methyl-2-pentanone, 2-methylpentanoic acid, 2-pentanone, 5-methyl-2-hexanone, 2,3-dichloro-2-methylbutane, 3-methyl-2-butanone, 2-propenilesther acetic acid and methanotiol

- Cosmetic formulations, PCPs and industrial uses: ethylbencene, dietilphthalate, 1,2,4-trimethylbenzene, methyl-isobutylcetone, eucaliptol, tetradecane, nonadecane.

Estrogenic hormones such as estrone, 17-β-estradiol, 17-β-estradiol acetate and 17-α-ethinylestradiol, were present in both residual waters and supply sources in ultra-trace amounts (ng/L). The highest concentrations were found in the residual water from Mexico City and in the Xotho canal (Table 5 and Table 6). The concentration of hormones in AR00 –Emisor Central and AR003-El Salto and AR004-Salado rivers, AR013-Xotho canal and in 019-Chilcuautla well is notable.

Supply source	Concentration (ng/L)	Supply source	Concentration (ng/L)
002 "Tezoquipa" Well	3.021	003 "El Tablón" Well	0.524
004 "Dendhó" Well	1.228	005 "San Antonio" Well	5.633
006 "El Refugio" Well	1.320	007 " Ajacuba" Well	2.105
008 "Noria Tetepango" Well	ND	009 "Doxey" Well	ND
010 "Tlaxcoapan" Well	0.134	011 "Teltipan" Well	3.105
012 "El Puedhe" Well	3.080	013 "Atengo" Well	ND
014 "Xochitlan" Well	0.158	015 "Progreso" Spring"	ND
016 "Progreso"	1.176	017 "Cerro Colorado" Spring	5.660
018 "Fitzhi" Well	2.110	019 "Chilcuautla" Well	11.426
020 "Pedro Ma. Anaya" Well	ND	023 "El Llano" Well	4.510

Table 5. Total estrogens in water supply sources (ND: no detected).

Sampling site	Concentration (ng/L)	Sampling site	Concentration (ng/L)
W "002 Emisor Central" Pipe/ canal	86.032	W 003 "El Salto" River	68.725
W 004 "Salado" River	ND	W 005 "Tlamaco Juandhó" Canal	ND
W 006 "Salto-Tlamaco 1" Canal	76.314	W 007 "Salto-Tlamaco 2" Canal	39.13
W 008 "Principal Requena 1" Canal	11.74	W 009 "Principal Dendhó" Canal	114.599
W 010 "Principal Requena 2" Canal	60.067	W 012"Principal Requena 3" Canal	33.855
W 013 "Xotho" Canal	12.141		

Table 6. Total estrogens in wastewater (ND: not detected).

In the second monitoring campaign six of the eight supply sources showed contamination from Phs and PCPs (Table 8), in ultra-trace concentrations: erythromycin, sulfamethoxazole (antibiotics), carbamazepine (antiepileptic), methylprednisolone (steroid anti-inflammatory used in cases of allergy), DEET (PCP), caffeine and benzoilecgonine (metabolite of cocaine). Chlofibric acid was found in 002-Tezoquipa (55.5 ng/L) and 003-El Tablón (36.67 ng/L), and AR003-El Salto (41.32 ng/L). The same river also showed traces of gemfibrozil (63.2 ng/L).

Source	Caffeine	Benzoyl-ecgonine	DEET	Methyl-prednisolone	Carba-mazepine	Erithromycin-H_2O	Sulfa-methoxasole
002	ND	ND	0.948	12.9	ND	ND	7.25
006	ND	ND	1.64	ND	ND	0.325	ND
013	24.3	4.93	2.01	ND	11.5	1.08	15.9
014	22.9	ND	1.01	ND	ND	0.31	16.7
017	ND	0.318	0.923	5.2	5.24	1.35	10.4
019	ND	ND	1.14	6.59	ND	1.05	22.8

Table 7. Pharmaceuticals and personal care products (concentration in ng/L; ND: not detected).

3.7. Bisphenol A, alkylphenols and alkylphenol-ethoxylates

As previously noted, bisphenol A and alkylphenols, including their ethoxylates, are relevant from a public health standpoint for their potential to alter endocrine processes in humans and aquatic and terrestrial organisms of various taxonomy. They are found in a variety of products of frequent use and are used as an adjuvant in the application of pesticides, which explains why, in the monitoring campaign that took place during the dry season, the highest concentrations were found in residual waters of AR02-Emisor Central and in AR03-El Salto y AR04-Salado rivers, as well as in the irrigation canals in the south zone. Furthermore, the concentrations of these contaminants in the supply sources were not homogenous and there was no obvious trend that suggested that wastewater from Mexico City was the only source of contamination. This is why we concluded that there was an influence from the interior of the Valley (Table 8. and Table 9).

Supply source	Concentration (µg/L)	Supply source	Concentration (µg/L)
002 "Tezoquipa" Well	ND	003 "El Tablón" Well	ND
004 "Dendhó" Well	ND	005 "San Antonio" Well	2.29
006 "El Refugio" Well	ND	007 " Ajacuba" Well	4.23
008 "Noria Tetepango" Well	2.10	009 "Doxey" Well	6.07
010 "Tlaxcoapan" Well	3.7	011 "Teltipan" Well	2.51
012 "El Puedhe" Well	0.18	013 "Atengo" Well	ND
014 "Xochitlan" Well	2.0	015 "Progreso" Spring"	ND
016 "Progreso"	ND	017 "Cerro Colorado" Spring	ND
018 "Fitzhi" Well	0.18	019 "Chilcuautla" Well	1.12
020 "Pedro Ma. Anaya" Well	ND	023 "El Llano" Well	ND

Table 8. Total alkylphenols and etoxilates in water supply sources (ND: not detected).

Sampling site	Concentration (µg/L)	Sampling site	Concentration (µg/L)
W "002 Emisor Central" Pipe/canal	41.28	W 003 "El Salto" River	64.57
WW 004 "Salado" River	17.41	W 005 "Tlamaco Juandhó" Canal	40.73
W 006 "Salto-Tlamaco 1" Canal	19.03	W 007 "Salto-Tlamaco 2" Canal	24.29
W 008 "Principal Requena 1" Canal	43.82	W 009 "Principal Dendhó" Canal	94.87
W 010 "Principal Requena 2" Canal	23.84	W 012"Principal Requena 3" Canal	25.9
W013 "Xotho" Canal	1.52		

Table 9. Total alkylphenols and etoxilates in wastewater.

3.8. Sanitary risks to the Mezquital Valley

Fifty-nine illnesses were identified as the primary causes of general mortality. Diabetes mellitus type II is at the top of the list, and ischemic illnesses of the heart, cerebral vascular ailments, diverse types of cancer (mainly breast, prostate, liver and leukemia) are among the primary causes of death. These diseases are multi-causal and related in great part to diet and life style as well as genetic and epigenetic factors; however environmental exposure to organic and inorganic contaminants can also contribute to the risk of contracting these ailments, which is why it is necessary to conduct studies that allows discovery or estimation of the contribution of water to these diseases.

For both samplings, hazard of exposure to 173 ECs was qualitatively identified, of which 35 are of high priority for their carcinogenic potential or systemic toxicity and among these, for their potential as EDs (acknowledged and suspicious [103-106]) and the frequency with which they were found of note are:

- Cosmetics, PCP and hygienic products formulations: ethylbenzene, octadecanoic acid (stearic acid), octyl-dimethyl-paramino benzoate (Escalol 507) and DEET

- Pharmaceuticals: carbamazepine, methylprednisolone, erythromicine

- Pesticides: 1,4-diclorobenzene

- Industrial uses: 1,2-dichloropropane, benzene, chloroform, n-propylbenzene, tetrahydro-furane, tetrachloroethylene, 1,1,2-trichloroethane

- Plasticizers: di-isobuthylphthalate, dipropylphthalate

- Estrogens: estrone, 17-β-estradiol, 17-β-estradiol acetate and 17-α-ethinylestradiol

By using a methodological approach of health risk assessment and analyzing the results of a survey completed by [107] from 1000 permanent residents in the Mezquital Valley, this study estimated the exposure to contaminants identified as hazardous through the use and consumption of water. The ingested dose via this route was compared to values of toxicity available on The Risk Assessment Information System data bases [108]. The hazard coefficient was obtained for arsenic, fluoride, bisphenol A, nonilphenol, naphthalene, and esters of ftalic acid or ftalates. The hazard coefficient or hazard index (HI) were estimated based on the volume of water consumed, concentration of the analyte in the water, intensity, frequency and magnitude of the oral exposure, body weight and the reference dose for critical effect in the target organ. A HI of 1.0 or less indicates that the exposure poses no health risks.

Hazard Index greater than 1.0 were identified for arsenic, mercury and fluoride, which means there are health risks associated with such substances in drinking water (Table 10). When such indexes are calculated for municipalities, greater hazards are identified. For example: Arsenic HI in Atitalaquia was 7.87 (± 5.08), in Progreso 2.29 (± 1.13) and in Tlaxcoapan 5.23 (± 2.47); Fluorie HI in Atitalaquia was 3.31 (± 2.14), in Atotonilco de Tula 2.98 (± 1.65), in Chilcuatla 4.68 (± 2.65), in Ixmiquilpan 5.40 (± 2.81), in Progreso 4.26 (± 2.33) and in Tlaxcoapan 3.03 (± 1.43).

Analyte	Media HI	Standard deviation	Min HI	Max HI
Arsenic	2.48E+00	4.07E+00	8.22E-03	2.41E+01
Mercury	3.26E-01	4.39E-01	0.00E+00	2.61E+00
Fluoride	3.95E+00	3.28E+00	2.74E-01	2.10E+01
Bisphenol A	7.18E-04	2.12E-03	0.00E+00	1.86E-02
4-Nonylphenol	2.81E-03	3.93E-03	0.00E+00	2.04E-02
Bis-2-(ethylhexyl)-phthalate	9.28E-03	3.51E-02	0.00E+00	3.07E-01
Diethylphthalate	5.78E-06	1.39E-05	0.00E+00	1.09E-04
Dibuthylphthalate	8.94E-04	3.04E-03	0.00E+00	2.68E-02

Table 10. Average Hazard Index from consumption of water with EDs (recognized and suspicious) within Mezquital Valley.

Even though at present it is not possible to associate health risks with chronic ingestion of organic EDs, there is uncertainty about exposure to small doses and the effects in the middle and long terms, besides the concomitant exposure.

3.9. Water treatment test

Pilot scale testing was carried out to make sure NF was effective at removing certain minerals, organic matter and ECs. Two supply sources within the valley were selected for this purpose.

The first phase of testing was conducted in the Cerro Colorado spring for 82 consecutive days. This source supplies water to 4 municipalities within the Valley (Mixquiahuala, Progreso, Tepacatepec and Tezontepec). Water characteristics are shown in Table 11. Due to these mineral characteristics of the water (hardness and silica content) up to 72% water recovery can be obtained.

Parameter	Concentration	Unit	Parameter	Concentration	Unit
Cl^-	148	mg/L	Na	195.14	mg/L
F^-	1.64	mg/L	Hardness	237.1	mg $CaCO_3$/L
$HCO_3{}^{2-}$	271.53	mg $CaCO_3$/L	SDT	1,147	mg/L
SO_4	169.96	mg/L	pH	7.16	
SiO_2	63.1	mg/L			

Table 11. Cerro Colorado Spring water quality.

Pilot scale NF tests were performed with a two stage skid; a 4:2 pressure vessel array with each pressure vessel holding 3 standard elements (NE4040-90 CSM), respectively, was utilized. Through brine recirculation, this design can simulate a two stage full-scale skid, employing six (8″ X 40″) elements per vessel.

In Figure 6 the pilot installation is shown: The whole treatment train includes sand filtration, a 5 m³ filtered water storage tank, 10 μm cartridge filtration, antifouling agent addition and NF.

The second phase of testing was conducted in the Tezoquipa well, located in Atitalaquia, Hgo. It lasted 36 days. In the first monitoring campaign, we found that the concentration of dissolved solids in this well was among the lowest in the Valley. However, heavy metals were found in the water, which is why it was selected for the second phase of the water treatment test.

Figure 6. Pilot unit including multimedia filtration and NF.

3.9.1. Rejection of mineral and organic compounds

To assess the performance of the NF pilot, six parameters were measured in feed water, permeate and brine: Conductivity, SO_4^{2-}, F^-, Hardness, HCO_3^- and TOC. Figures 7, 8 and 9 show the rejection of those parameters monitored during water treatment tests.

Mineral content rejection, at the beginning of water treatment testing, was 80% (measuring conductivity). This level of efficiency was about 10% below the rejection calculated with the CSM4PRO design software (corresponding to membranes NE4040-90 of CSM). However, as the test progressed, the salt rejection rose to 93%, which was very close to the calculations made by CSM4PRO.

HCO_3^- removal was 92% on average (± 0.9981) while hardness was rejected at a higher rate of 97.46% on average, with a minimum of 95.12 and a maximum of 98.95%.

Sulphate removal was 98.78% on average, with a maximum of 99.57% and a minimum of 95.75%. These results are consistent with the characteristics of the membrane, since it is a negatively charged hydrophobic membrane specialized in the removal of polyvalent anions.

Fluoride rejection oscillated between 80% and 96.77% (Figure 8). The variability in the results could be explained by changes in feed water temperature. Filtered water coming from the spring was stored in the 5m³ tank. The tank was exposed directly to sunlight, causing a rise in water temperature. While samples and measurements were meant to be taken at 13:00 hours, this was not always possible and there was no control over ambient temperatures.

Organic matter, measured as TOC, was rejected at a rate of 82% on average (Figure 9). Nevertheless, its rejection rate varied from 41% to 100%. It is worth mentioning that in two of the four measurements in which removal efficiency was less than 45%, the concentration of TOC in the raw water was lower than 2 ppm; even so the variability is considerable.

With respect to pharmaceuticals and personal care products, laboratory data indicates that the membranes were capable of rejecting all of the sulfamethoxazole% (Table 12).

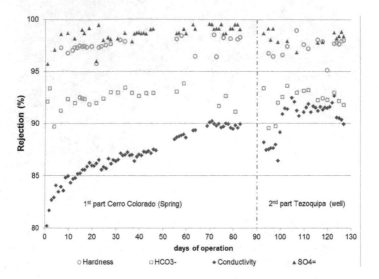

Figure 7. Major ions rejection observed in water tests.

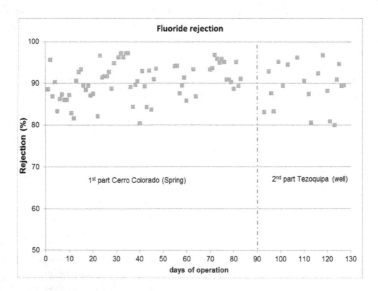

Figure 8. Fluoride removal efficency observed in water treatment test.

Erythromycin-H$_2$O and cocaine were found in the brine but not in the feed water or permeate. Given the low concentrations reported (fractions of nanograms), it is plausible that the raw water contained these compounds at non-detectable levels and that they were rejected with great efficiency, causing these substances to reach concentrations at a detectable and quantifiable level.

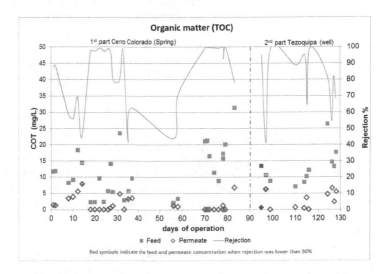

Figure 9. NF performance for organic matter removal.

Site	Compound	Concentration (ng/L)
Raw water	Sulfamethoxazole	7.25
	DEET	0.948
	Metylprednizolone	12.9
Permeate	DEET	2.01
	Metylprednizolone	11
Brine	Caffeine	14.5
	Eritromicina-H2O	0.654
	Sulfamethoxazole	18.8
	Cocaine	0.172
	DEET	0.848

Table 12. Pharmaceuticals and PCPs in water treatment test.

The NF membranes do not appear to be effective in the removal of DEET and are inefficient in the removal of methylprednisolone. The molecular size of DEET is smaller than the

Molecular Weight Cut-Off of the membrane. Also, DEET is a neutral molecule, so negatively charged membranes won't exert electrostatic repulsion. In the case of methylprednisolone, it is of note that it was not found in the brine. This, according to the mass balance, is not possible. In other words, if the concentration in the feed and the permeate streams are the same, the same should be true for the brine.

Overall, groundwater quality within the Valley does not meet the Mexican Drinking Water Quality Standard [94], from a total dissolved solids standpoint [94]. In some sources hardness, sulphate, fluoride and heavy metals like lead, mercury and arsenic are present at concentrations exceeding the MCL. NF was able to reject all these substances at such a rate that the permeate met the Mexican Drinking Water Quality Standard. From the mineral rejection standpoint, NF seems to be an interesting alternative to treat the Mezquital Valley ground water in order to produce drinking water.

4. Conclusions and research needs

In the semi-arid Mezquital Valley, reusing wastewater for agricultural purposes has benefitted aquifer recharge, which has increased the availability of water destined for various human activities, such as drinking water. Nevertheless, the presence of As, F, Hg and Pb and multiple organic pollutants in 19 water sources that were sampled showed a wide variety of potential health risks, including disruptions of the endocrine system.

The Hazard Quotients (HQ) estimated for the EDs (acknowledged and suspected) in the mentioned area indicated risks associated (HC > 1.0) with As and F, while no risks were found at the time of the study for organic EDs (phtalates, nonilphenols and bisphenol A). Nevertheless, in water reuse scenarios such as this one, even with the limitations inherent to both point monitoring and considering the lack of scientific knowledge.

It is necessary to emphasize the prevention of human exposure to EDs, considering that: a) the effects of critical exposure during intrauterine, perinatal and puberty periods may not manifest until adulthood; b) the population in the Mezquital Valley is exposed to a mixture of EDs that could lead to the addition, synergy, potentiation or antagonism of effects, and c) it may be assumed that the entire population is potentially exposed to these pollutants since, in qualitative terms, all water sources presented at least one ED.

More monitoring campaigns focusing on EDs in a greater number of supply sources are necessary to quantify the EDs with greater precision and estimate exposure with higher accuracy.

Studies are required to integrate environmental aspects, life style and toxicological research with epidemiology (cohort retrospective study or control cases) that allow associations to be established between EDs exposure with ailments or specific health issues: i.e. diabetes mellitus type II, the development of various types of cancer, reproductive function and processes and neurodevelopment, which are currently of interest to various groups of researchers and the governmental agencies responsible for regulation and policies in health, water and the environment.

On the other hand, the byproducts of the degradation and transformation of a compound can also alter the hormonal system and, while a large number of researchers believe that physico-chemical characteristics of the molecules determine the molecules' behavior and environmental distribution, a greater understanding of their action, transformation and environmental fate are also required to assess and minimize the health risks associated with exposure to EDs in concentrations relevant to drinking water.

Water delivered through public networks must be treated in order to comply with the Mexican Standard for water quality. A NF process will remove mineral pollutants found in the Mezquital Valley Nevertheless, from an ED standpoint, further tests are needed in order to obtain enough evidence to determine if NF is capable of producing drinking water from these sources, or another process is needed, either as a substitution (like RO) or a complement (like activated carbon).

Acknowledgements

Projects Coordinating Office for the Valley of Mezquital /National Water Commission.

Author details

J. E. Cortés Muñoz[1*], C. G. Calderón Mólgora[1], A. Martín Domínguez[1], E. E. Espino de la O[2], S. L. Gelover Santiago[1], C. L. Hernández Martínez[2] and G. E. Moeller Chávez[1]

*Address all correspondence to: jucortes@tlaloc.imta.mx

1 Mexican Institute of Water Technology, Mexico

2 National Water Commission, Mexico

References

[1] Arnold, B, & Calford, J. Treating water with chlorine at point of-use to improve water quality and reduce child diarrhea in developing countries: a systematic review and meta-analysis. American Journal of Tropical Medicine and Hygiene (2007). , 76(2), 354-64.

[2] Cairncross, S, Hun, C, Boisson, S, Bostoen, K, Curtis, V, Fung, I, & Schmidt, W. Water, sanitation and hygiene for the prevention of diarrhoea. International Journal of Epidemiology (2010). ii205., 193.

[3]   Haque, R, Guha, D. N, Samanta, S, Ghosh, N, Kalman, D, Smith, M, Mitra, S, Santra, A, Lahri, S, & Das, S. De B, Smith A. Arsenic water and skin lesions: dose-response data from West Bengal, India. Epidemiology (2003). , 14, 174-82.

[4]   Working Group on the Evaluation of the Carcinogenic Risks to HumansArsenic in drinking water. In: IARC Monographs Some drinking-water disinfectants and contaminants, including arsenic. Lyon: WHO (2004). , 84, 39-270.

[5]   Ayoob, S, & Gupta, A. Fluoride in drinking water: a review on the status and stress effects. Critical Reviews in Environmental Science and Technology (2006). , 36, 433-87.

[6]   Bradley, P. Potential for biodegradation of contaminants of emerging concern in stream systems. Conference Proceedings, October 14-15, (2008). Charleston Area Event Center, South Carolina, USA.

[7]   Mastroianni, N, López, M, & Barceló, D. Emerging contaminants in aquatic environments: state of the art and recent scientific contributions. Contributions to Science (2010). , 6(2), 193-97.

[8]   Damstra, T, Barlow, S, Bergman, A, & Kavlock, R. van Der Kraak G editors. Global Assessment of the State of the Science of endocrine Disruptors. USA:WHO/IPCS; (2002).

[9]   Brausch, J, & Rand, G. A review of personal care products in the aquatic environment: Environmental concentrations and toxicity. Chemosphera (2011). , 82-1518.

[10] Santos, L, Araújo, A, Fachini, A, Pena, A, Delerue-matus, E, & Montenegro, M. Ecotoxicological aspects related to the presence of pharmaceuticals in the aquatic environment. Journal of Hazardous Materials (2010). , 175-45.

[11] Casals, C, & Desvergne, C. Endocrine disruptors: from endocrine to metabolic disruption. Annu. Rev. Physiol(2011). , 73, 135-62.

[12] Salame, A, Méndez, F, Aguirre, G, & Serrano, H. Disrupción endocrina de la diferenciación sexual. ContactoS (2008). , 70, 43-9.

[13] Brevini, T, Bertola, S, & Zanetto, C. Effects of endocrine disruptors on developmental and reproductive functions. Current Drug Targets-Immune, Endocrine & metabolic disorders (2005). , 5-1.

[14] Snyder, S, Villeneuve, D, Snyder, E, & Giesy, J. Identification and quantification of estrogen receptor agonists in wastewater effluents.Environ. Sci. Technol (2001). , 35, 3620-25.

[15] Barnes, K, Kolpin, D, Furlong, E, Zaugg, S, Meyer, M, & Barber, L. A national reconnaissance of pharmaceuticals and other organic wastewater contaminants in the United States- I) Groundwater. Science of the Total Environment (2008). , 402, 192-200.

[16] Stackelberg, P, Furlong, E, Meyer, M, Zaugg, S, & Henderson, . . Persistence of pharmaceutical compounds and other organic wastewater contaminants in conventional drinking water treatment plant. Science of the Total Environment 2004;329:99-113.

[17] The Prague Declaration on Endocrine Disruptionhttp://www.bio.uni-frankfurt.de/ee/
ecotox/news/PragueDeclaration.pdfaccessed 15 may (2012).

[18] Rahman, M, Yanful, E, & Jasim, S. Endocrine disrupting compounds (EDCs) and
pharmaceuticals and personal care products in the aquatic environment: implications
for the drinking water industry and global environmental health. Journal of Water and
Health (2009). doi:10.2166/wh.2009.021., 2, 224-43.

[19] Chemcial Abstracts Servicehttp://www.cas.org/index.htmlaccessed 16 may (2012).

[20] Acerini, C, & Hughes, I. Endocrine disrupting chemicals: a new and emerging public
health problem? Arch dis Child (2006). , 91(8), 633-38.

[21] Benachour, N, Clair, E, Mesnage, R, & Séralini, G. endocrine disruptors: new discov-
eries and possible progress of evaluation. In: Berhardt V. (ed.) Advances in Medicine
and Biology. France: Nova Science Publishers; (2012). , 1-57.

[22] Epa, U. S. Special report on Environmental Endocrine Disruption: An Effects Assessment and Analysis. Office of Research and Development (1997). EPA/630/R-96/012,
Washington D.C.

[23] López, J, & Barceló, D. Contaminantes orgánicos emergentes en aguas continentales y
aspectos relacionados con el marco normativo y planificación hidrológica en España.
En: Aguas continentales. Madrid:, 91-111.

[24] Burkhardt-holm, P. Endocrine disruptors and water quality: a state of the art review.
Water Resources Development (2010). , 26(3), 477-93.

[25] Diamanti, E, Bourguignon, J. P, Giudice, L, Hauser, R, Prins, G, Soto, A, Zoeller, T, &
Gore, A. Endocrine-disrupting chemicals: an Endocrine Society Scientific statement.
Endocrine Reviews (2009). , 30, 293-342.

[26] Blasco, J. Del Valls A. Impact of emergent contaminants in the environment: Environmental risk assessment.Hdb Env Chem. (2008). Part S/1):169-88.

[27] Nikolaou, A, Meric, S, & Fatta, D. Occurrence patterns of pharmaceuticals in water and
wastewater environments. Anal Bioanal Chem (2007). , 387, 1225-34.

[28] Godfrey, E, Woessner, W, & Benotti, J. Pharmaceuticals in on-site sewage effluent and
ground water, Western Montana. Groundwater (2008). , 45(3), 263-71.

[29] Asano, T, & Cotruvo, J. Groundwater recharge with reclaimed municipal wastewater:
health and regulatory considerations. Water Res. (2004). Bound J, Voulvoulis N.
Household disposal of pharmaceuticals as a pathway for aquatic contamination in the
United Kingdom. Environ Health Perspect 2005;113:1705-11., 38, 1941-51.

[30] Vethaak, D, Lahr, J, Schrap, S, Belfroid, A, Rijs, G, Gerritsen, A, Boer, J, Bulder, A,
Grinwis, G, Kuiper, R, Legler, J, Murk, T, Peijnenburg, W, Verhaar, H, & Voogt, P. An
integrated assessment of estrogenic contamination and biological effects in the aquatic
environment of The Netherlands. Chemosphere (2005). , 59, 511-24.

[31] SirbuD, Curseu D, Popa M, Achimas A, Moldovan Z. Environmental risks of pharma-
     ceuticals and personal care products in water.Tenth International Water Technology
     Conference, IWTC10 2006, Alexandria, Egypt. http://www.iwtc.info/2006_pdf/15-2.pdf
     (accessed 15may 2012).

[32] Daughton, C, & Ternes, T. Pharmaceuticals and personal care products in the environ-
     ment: Agents of subtle change?Environ. Health Perspect. (1999). suppl. 6):907-38.

[33] Trussell, R. Constituents of Emerging Concern: An Overview In: WEFTEC (2006).
     Session 20 of Proceedings of the Water Environment Federation, October, WEFTEC
     October 2006, Dallas Tx, USA.http://www.environmental-expert.comaccessed 15 may
     2012)., 2006, 21-25.

[34] Snyder, S, Trenholm, R, Snyder, E, Bruce, G, Pleus, R, & Hemming, J. Toxicological
     relevance of EDCs and pharmaceuticals in drinking water. USA: AWWA Research
     Fundation/American Water Works Association/IWA; (2008).

[35] Committee on Hormonally Active Agents in the EnvironmentNational Research
     Council. Hormonally active agents in the environment. USA: National Academies
     Press; (1999).

[36] Snyder, S, Westerhoff, P, Yoon, Y, & Sedlak, D. Pharmaceuticals, personal care
     products, and endocrine disruptors in water: implications for the water industry.
     Environmental Engineering Science (2003). , 20(5), 449-69.

[37] Hanselman, T, Graetz, D, & Wilkie, A. Manure-borne estrogens as potential environ-
     mental contaminants: a review. Environmental Science & Technology (2003). , 37(24),
     5471-78.

[38] Rivas, A, Granada, A, Jiménez, M, & Olea, N. Exposición humana a disruptores
     endocrinos. Ecosistemas (2004). , 13(3), 7-12.

[39] Snyder, E, Snyder, R, Pleus, R, & Snyder, S. Pharmaceuticals and EDCS in the US wáter
     industry-an update. Journal AWWA (2005). , 97(11), 32-5.

[40] Schug, T, Janesick, A, Blumberg, B, & Heindel, J. Endocrine disrupting chemicals and
     disease susceptibility. Journal of Steroid Biochemistry & Molecular Biology (2011). ,
     127, 204-15.

[41] Olujimi, O, Fatoki, O, Ordendaal, J, & Okonkwo, O. Endocrine disrupting chemicals
     (phenol and phthalates) in the South African environment: a need for more monitoring.
     Water SSA (2010). , 36(5), 671-81.

[42] Australian Academy of ScienceEndocrine disruption. Are chemicals in our environ-
     ment reducing male fertility and causing cáncer? Do they cause deformities and effect
     reproduction in wildlife? Australia:Ntional Science and Industry Forum Report; (1998).

[43] Argemi, F, Cianni, N, & Porta, A. Disrupción endocrina: perspectivas ambientales y
     salud pública. Acta Bioquím Clín Latinoam (2005). , 39(3), 291-300.

[44] Chichizola, C. Disruptores endocrinos. Efectos en la reproducción. Revista Argentina de Endocrinología (2004). , 41(2), 78-105.

[45] Laws, S, Yavanzhay, S, Cooper, R, & Eldridge, J. Nature of the binding interaction for 50 structurally diverse chemicals with rat estrogen receptors. Toxicological Sciences (2006). , 74(1), 46-56.

[46] SandersonThe steroid hormone biosynthesis pathway as a target for endocrine-disrupting chemicals. Toxicological Sciences (2006). , 3-21.

[47] Tabb, M, & Blumberg, B. New modes of action for endocrine-disrupting chemicals. Molecular Endocrinology (2004). , 20(3), 475-82.

[48] Chichizola, C, Scaglia, H, Franconi, C, Ludueña, B, Mastandrea, C, & Ghione, A. Disruptores endocrinos y el sistema reproductive. Revista Bioquímica y Patología Clínica (2009). , 73(3), 9-23.

[49] Chichizola, C. Disruptore endocrinos. Efectos en la reproducción. Revista Argentina de endocrinología y Metabolismo (2003). , 40(3), 172-88.

[50] Swedenborg, E, Rüegg, J, Mäkelä, S, & Pontgratz, I. Endocrine disruptive chemicals: mechanisms of action and invelovement in metabolic disorders. Journal of Molecular endocrinology (2009). , 43, 1-10.

[51] Welshons, W, Thayer, K, Judy, B, Taylor, J, & Curran, E. Vom Saal F. Large effects from small exposures. I. Mechanisms for endocrine-disruptingchemicals with estrogenic activity. Environ Health Perspect (2003). , 111, 994-1006.

[52] Vom Saal FHughes C. An extensive new literature concerning low-dose effects of Bisphenol A shows the need for a new risk assessment. Environ Health Perspect (2005). , 113, 926-33.

[53] Kortenkamp, A, Evans, R, Martin, O, Mckinlay, R, Orton, F, & Rosivatz, E. Annex 1 Summary of the state of the science. In: State of the art assessment of endocrine disrupters. Final report. Anex 1. European Commission, DG Environment;(2012).

[54] Okeda, H, Tokunaga, T, Liu, x, Takayanagi, S, Matsushima, A, & Shimohigashi, Y. Direct evidence revealing structural elements essential for the high binding ability of Bisphenol A to human estrogen-related Receptor-γ. Environ Health Perspect (2007). , 116, 32-8.

[55] Melzer, D, Harries, L, Cipelli, R, Henley, W, Money, C, Mccormack, P, Young, a, Guralnik, J, Ferrucci, L, Bandinelli, S, Corsi, A. M, & Galloway, T. Bisphenol A exposure is associate in vivo estrogenic gene expression in adults. Environ Health Perspect (2011). , 119, 1788-93.

[56] Bourguignon, J, & Parent, A. Early homeostatic disturbances of human growth and maturation by endocrine disrupters. Current Opinion in Pediatrics (2010). , 22, 470-77.

[57] Frye, C, Bo, E, Calamadrei, G, Calzà, L, Dessì-fulgeri, F, Fernández, M, Fusani, L, Kah, O, & Kajta, M. Le Page G, Venerosi A, Wojtowicz A, Panzica G. Endocrine disrupters:

a review of some sources, effects, and mechanisms of action on behaviour and neuro-endocrine systems. Journal of Endrocrinology (2011). doi:j.x., 1365-2826.

[58] Rees, E, Todd, M, Beam, J, & Aiello, A. The Impact of Bisphenol A and triclosan on immune parameters in the U.S. population, NHANES (2003). Environ Health Perspect. 2011;, 119(3), 390-96.

[59] Braun, J, Kalkbrenner, A, Calafat, A, Yolton, C, Ye, X, Dietrich, K, & Lanphear, B. Impact of early-life Bisphenol A exposure on behavior and executive function in children. Pediatrics (2011). , 128(5), 883-81.

[60] Guerrero, C, Settles, M, Lucker, B, & Skinner, K. Epigenetic transgenerational actions of vinclozolin on promoter regions of the sperm epigenome (2010). PLoS ONE 5(9):e13100. doi:10.1371/journal.pone.0013100.

[61] Palmer, J, Herbst, A, Noller, K, Boggs, D, Troisi, R, Titus, L, Hatch, E, Wise, L, Strohs-nitter, C, & Hoover, R. Urogenital abnormalities in men exposed to diethylstilbestrol in utero: a cohort study. Environmental Health (2009). doi:10.1186/X-8-37

[62] Ingerslev, F, Vaclavick, E, & Halling-sorensen, B. Pharmaceuticals and personal care products: A source of endocrine disruption in the environment?.Pure Appl. Chem. (2003).

[63] Su, T, Pagliaro, M, Schmidt, T, Pickar, D, Wolkowitz, O, & Rubinow, D. Neuropsychi-atric effects of anabolic steroids in male normal volunteers. JAMA (1993). , 269(21), 2760-64.

[64] Assessment of Technologies for the Removal of Pharmaceuticals and Personal Care Products in Sewage and Drinking Water Facilities to Improve the Indirect Potable Water Reuse: POSEIDONhttp//wwweu-poseidon.com (accessed 2 August (2011).

[65] Yoon, Y, Westerhoff, P, Snyder, S. A, Wert, E. C, & Yoon, J. Removal of endocrine disrupting compounds and pharmaceuticals by nanofiltration and ultrafiltration membranes. Desalination (2006). , 202-16.

[66] Bellona, C, & Drewes, J. E. The role of membrane surface charge and solute physico-chemical properties in the rejection of organic acids by NF membranes. Journal of Membrane Science (2005).

[67] Drewes, J. E, Bellona, C, Luna, J, Hoppe, C, Amy, G, Filteau, G, Oelker, G, Lee, N, Bender, J, & Nagel, R. Can nanofiltration and ultra-low pressure reverse osmosis membranes replace RO for the removal of organic micropollutants, nutrients and bulk organic carbon?- A pilot-scale investigation. In: WEFTEC (2005). October- 2 November 2005, Washington, DC, USA Wef Proceedings; 2005.

[68] Xu, P, Drewes, J. E, Bellona, C, Amy, G, Kim, T, Adam, U, & Heberer, M. T. Rejection of Emerging Organic Micropollutants in Nanofiltration-Reverse Osmosis Membrane Applications. Water Environment Research (2005). , 77(1), 40-48.

[69] Bellona, C, & Drewes, J. Viability of a low-pressure nanofilter in treating recycled water for water reuse applications: A pilot-scale study. Water Research (2007). , 41-348.

[70] Park, C, Jang, H, Yoon, Y, Hong, S, Jung, J, & Chung, Y. Removal characteristics of endocrine disrupting compounds and pharmaceuticals by NF membranes. In: 4th IWA International Membrane Conference, Membranes for Water and Wastewater Treatment, May (2007). Harrogate, UK. IWA; 2007, 15-17.

[71] Lee, S, Quyet, N, Lee, E, Kim, S, Lee, S, Jung, Y, Choi, S.H, & Cho, J. . Efficient removals of tris(2-chloroethyl) phosphate (TCEP) and perchlorate using NF membrane filtrations. Desalination 2008; 221 234-237.

[72] Koyuncu, I, Arikan, O. A, Wiesner, M. R, & Rice, C. Removal of hormones and antibiotics by nanofiltration membranes. Journal of Membrane Science (2008). , 309-94.

[73] Radjenovic, J, Petrovic, M, Ventura, F, & Barceló, D. Rejection of pharmaceuticals in nanofiltration and reverse osmosis membrane drinking water treatment. Water Research (2008). , 42-3601.

[74] Schäfer, A. I, Fane, A. G, & Waite, T. D. Editors. Nanofiltration- Principles and Applications. Elsevier; (2005).

[75] Bellona, C, Drewes, J, Xu, P, & Amy, G. Factors affecting the rejection of organic solutes during NF/RO treatment-a literature review. Water Research (2004). , 38-2795.

[76] Agenson, K. O, Oh, J. I, & Urase, T. Influence of molecular structure on the rejection characteristics of Volatile and Semi-volatile Organic Compounds by Nanofiltration. In: IMSTEC'03 5th International Membrane Science and Technology Conference, November (2003). Australia. New South Wales: UNESCO Centre for Membrane Science and Technology 2003, 10-14.

[77] Zhang, Y, Causserand, C, Aimar, P, & Cravedi, J. P. Removal of bisphenol A by nanofiltration membrane in view of drinking water production. Water Research (2006). , 40-3793.

[78] Kimura, K, Amy, G, Drewes, J, Herberer, T, Kim, T. U, & Watanabe, Y. Rejection of organic micropollutants (disinfection by products, endocrine disrupting compounds, and pharmaceutically active compounds) by NF/RO membranes. Journal of Membrane Science (2003). , 227-113.

[79] Kiso, Y, Nishimura, Y, Kitao, T, & Nishimura, K. Rejection properties of non-phenolic pesticides with nanofiltration membranes. Journal of Membrane Science (2000). , 171-229.

[80] Nghiem, L. D, Schäfer, A. I, & Elimelech, M. Removal of Natural Hormones by Nanofiltration Membranes: Measurement, Modeling, and Mechanisms. Environmental Science Technology (2004). , 38(6), 1888-1896.

[81] Kim, J. H, Kwon, H, Lee, S, & Lee, C. H. Removal of endocrine disruptors using homogeneous metal catalyst combined with nanofiltration membrane. Water Science and Technology (2005).

[82] INEGIMéxico en cifras. Hidalgo. http://www.inegi.org.mx/sistemas/mexicocifrasacceded 28 September (2010).

[83] Gobierno del estado de Hidalgohttp://www.hidalgo.gob.mx/acceded 27 September (2010).

[84] Water National CommissionBritish Geological Survey, London School of Hygiene and Tropical Medicine and University of Birmingham. Impact of Wastewater Reuse onGroundwater in the Mezquital Valley, Hidalgo State, Mexico. Final Report Technical Report WC/98/42. (1998).

[85] Lesser, L, Lesser, J, Arellano, S, & González, D. Balance hídrico y calidad del agua subterránea en el acuífero del Valle del Mezquital, México central. Revista Mexicana de Ciencias Geológicas (2011). , 28(3), 323-36.

[86] Jiménez, B, & Chávez, A. Quality assessment of an aquifer recharged withwastewater for its potential use as drinking source:"El Mezquital Valley" case. Water Science and Technology (2004). , 50(2), 269-76.

[87] Downs, T, Cifuentes, E, Ruth, E, & Suffet, I. M. Effectiveness of natural treatment in a wastewater irrigarion district of the Mexico City: A synoptic field survey. Water Environment Research (2000). , 72(1), 4-21.

[88] Siebe, C, & Cifuentes, E. Environmental impact of wastewater irrigation in Central Mexico: An overview. Int. J. Environ. Health Res (1995). , 5, 161-73.

[89] Siemens, J, Huschek, G, Siebe, C, Kaupenjohann, d, & Concentrations, M. and mobility of human pharmaceuticals in the world's largest wastewater irrigation system, Mexico City-Mezquital Valley. Water Research (2007).

[90] Gibson, R, Becerril, E, Silva, V, & Jiménez, B. Determination of acidic pharmaceuticals and potential endocrine disrupting compounds in wastewaters and spring waters by selective elution and analysis by gas chromatography-mass spectrometry. Journal of Chromatography A (2007).

[91] AXYS Method MLA-075 Rev 03 Ver 02: AXYS Method MLA-075: Analytical procedures for the analysis of pharmaceutical and personal care products in solidaqueous and tissue samples by LC-MS/MS. Canada (2011).

[92] Method, U. S-E. P. A, & Purge, B. and Trap for aqueous samples.

[93] Method, U. S-E. P. A. D. Semivolatile Organic Compounds by Gas Chromatography / Mass Spectrometry (GC/MS).

[94] Secretaría de SaludMODIFICACION a la Norma Oficial Mexicana NOM-SSA1-(1994). Salud ambiental. Agua para uso y consumo humano. Límites permisibles de calidad y

tratamientos a que debe someterse el agua para su potabilización. Diario Oficial de la Federación 22 de noviembre de 2000. México., 127.

[95] Özen, S. Darcan Ş. Effects of environmental endocrine disruptors on pubertal development. J Clin Res Ped Endo (2011). DOI:jcrpe., 3i1

[96] Committee on Fluoride in Drinking WaterNational. Research Council. Fluoride in drinking water: A scientific review of EPA´s standards. USA: NAP; (2006).

[97] Davey, J, Bodwell, J, Gosse, J, & Hamilton, J. Arsenic as an Endocrine disruptor: effects of arsenic on estrogen receptor-mediated gene expression in vivo and in cell culture. Toxicological Sciences (2007). , 98(1), 75-86.

[98] Bodwell, J, Gosse, J, Nomikos, P, & Hamilton, J. Arsenic disruption of steroid receptor gene activation: complex dose-response effects are shared by several steroid receptors. Chem Res Toxicol (2008). , 19(12), 1619-29.

[99] Barr, F, Krohmer, L, Hamilton, J, & Sheldon, L. Disruption of histone modification and CARM1 recruitment by arsenic represses transcription at glucocorticoid receptor-regulated promoters. PLoS One (2009). e6766;doi:10.1371/journal.pone.0006766.

[100] Xiang, Q, Chen, L, Liang, Y, Wu, M, & Chen, B. Fluoride and thyroid function in children in two villages in China. Journal of Toxicology and Environmental Health Sciences (2009). , 1(3), 54-9.

[101] Susheela, A, Bhatnagar, M, Vig, K, & Mondal, N. excess fluoride ingestion and thyroid hormone derangements in children living in Delhi, India. Fluoride (2005)., 38(2), 98-108.

[102] Stein, J, Schettler, T, Wallinga, D, & Valenti, M. In harm's way: toxic threats to child development. J Dev Behav Pediatr (2002). SS22., 13.

[103] Walker, J, Fang, H, Perkins, R, & Tong, W. QSARs for endocrine disruption priority setting Database 2: The Integrated 4-Phase Model. Quant. Struct.-Act. Relat. (2003). , 22, 1-17.

[104] USEPAEPA's List of 134 Endocrine Disruptors. http://www.epa.gov/endo/pubs/ prioritysetting/list2facts.htmaccessed 02 June (2012).

[105] De Coster, s, & Van Lare, B. Endocrine disrupting chemicals, associated disorders and mechanisms of action.www.hindawi.com/journals/jeph/aip/713696.pdfaccessed02June (2012).

[106] Scorecard Pollution sitehttp://scorecard.goodguide.com/index.tclaccessed 16 February (2011).

[107] USEPARisk Assessment Guidance for Superfound.Vol. I. Human Health Evaluation Manual (Part A). USA:EPA;(1989).

[108] Risk Assessment Information SystemChemical Toxicity values. http://rais.ornl.gov/cgi-bin/tools/TOX_searchaccessed 16 February (2011).

Chapter 2

Bloom-Forming Cyanobacteria and Other Phytoplankton in Northern New Jersey Freshwater Bodies

Tin-Chun Chu and Matthew J. Rienzo

Additional information is available at the end of the chapter

1. Introduction

1.1. Phytoplankton

Phytoplankton not only plays a vast role in the aquatic food chain, but some groups are essential in the production of atmospheric oxygen [1]. Phytoplankton include cyanobacteria, algae and many other groups. Some of the most common types of phytoplankton in North American freshwater bodies include species of Bacillariophyceae (diatoms) as well as thousands of species of cyanobacteria.

Diatoms are a type of phytoplankton that possess several unique contours due to a cell wall composed of silicon dioxide (SiO_2) [2, 3]. The diatoms, or Bacillariophyta, have distinct structures and thus are easily identifiable in a water sample. Diatoms can be found in a large range of pH and dissolved oxygen values as well as in ecosystems with a wide concentration of solutes, nutrients, contaminants, and across a large range of water temperatures due to their durable cell walls [2].

There are many species of cyanobacteria, commonly found in freshwater lakes and ponds as well as marine environments. Originally called blue-green algae because of their color, cyanobacteria is a phylum of bacteria that uses photosynthesis to obtain energy. Cyanobacteria are prokaryotes and possess the pigment chlorophyll *a*, which is necessary for oxygenic photosynthesis and can be exploited during molecular analysis to detect the presence of cyanobacteria in a sample [4]. Cyanobacteria aided in the transformation of the Earth's atmosphere by producing atmospheric oxygen [1]. Freshwater cyanobacteria can be found as unicellular, filamentous, or colonial cells within the environment. Some of the common cyanobacteria found in freshwater sources in North America include Synechococcus, Anabaena, Oscillatoria, Nostoc, and Anacystis [2].

1.2. Algal blooms

Although the necessity for cyanobacteria and other phytoplankton in the environment is apparent, overgrowth in urbanized areas due to eutrophication results in formation of algal blooms, causing deleterious effects to both aquatic life as well as anything that may come in contact with the water. Some of the common algal bloom-forming cyanobacteria include those with filamentous and colonial cells [5].

Eutrophic freshwater ecosystems may contain a high average algal biomass include phytoplanktons such as cyanobacteria, chlorococcales or dinoflagellates [1, 6]. Eutrophication is the water body's response to added nutrients like phosphate and nitrates. In urbanized areas, the human factor of nutrient introduction to these ecosystems, otherwise known as cultural eutrophication, has recently been considered as one of the most important factors driving the increase in algal bloom frequency as well as intensity [7]. Fertilizer runoff, car washing, and pet wastes being discarded into storm drains are three major modern events causing changes that disturb existing equilibrium between phytoplankton and other aquatic life, accelerating eutrophication [1]. The algal mat that forms at the water's surface can easily prevent sun from penetrating the lower portions of the water. In figure 1 below, an extensive algal bloom is seen in Branch Brook State Park Lake in Newark, NJ.

1.3. Cyanotoxin

Algal bloom production can be harmful due to decreased sunlight penetration, decreased dissolved oxygen, and also possible toxin release by certain species of cyanobacteria [8]. Many species of cyanobacteria can produce toxins, posing a further risk for aquatic life. There are about 50 species of cyanobacteria that have been shown to produce toxins which are harmful to invertebrates. Microcystis, Anabaena, Oscillatoria, Aphanizomenon, and Nodularia are a few genera which contain species known to produce cyanotoxins. There are three main types of cyanotoxins. Neurotoxins affect the nervous system, hepatotoxins affect the liver, and dermatoxins affect the skin (NALMS) [9]. It could pose a serious threat for both human and animal health if they consume the water from the contaminated sites. Microcystins and other cyanotoxins are heat stable, thus cannot be destroyed by boiling. Also, many cyanotoxins are not easily separated from drinking water if they are dissolved in water. Currently, there are several cyanotoxins that are on the US EPA Contaminant Candidate List (CCL2) which are being evaluated for human toxicity (NALMS) [10]. Exposure routes of these cyanotoxins are dependent on the purpose of the contaminated water. If the contaminated water is part of a reservoir, the exposure route may be ingestion due to improperly filtered drinking water. If the contaminated water is used for recreational use, the exposure route may be skin, ingestion, or inhalation. Human exposure may also come from ingestion of animals that were living in the contaminated water. Saxitoxins, known neurotoxins secreted by several cyanobacterial species including *Anabaena circinalis*, are also known as paralytic shellfish toxins (PSTs). These neurotoxins infect shellfish, which in turn infect humans who ingest those shellfish [11]. This neurotoxin, along with the other cyanotoxins produced by cyanobacteria, currently has no antidote [12].

1.4. Detection and treatment of algal blooms

The need for treatment of contaminated freshwater across the world is at an all-time high due to the increase in urbanization. In order to prevent harmful algal blooms from forming, it is necessary to understand the balance between cyanobacteria and their viruses, cyanophage. Cyanophage are viruses that infect cyanobacteria in a species specific manner and are just as ubiquitous as cyanobacteria in ecosystems [13]. So, the first step in possibly detecting and preventing algal bloom formation is to identify the common species at each susceptible freshwater body. Microscopy can be used to identify organisms found within environmental samples. Microscopy allows for identification as well as determination of cell density within the sample. Microscopy, unfortunately, is inefficient and time consuming. As a complement to microscopy, polymerase chain reaction (PCR) can be employed. PCR can be used to prime for conserved regions among all phyla of cyanobacteria and other phytoplankton. In an environmental sample, it is important to first perform PCR using universal cyanobacterial primers in order to determine the presence of cyanobacteria. There have been previous studies in which both universal and phyto-specific primers have been determined to be effective in amplifying the 16s rRNA genes in cyanobacteria [14, 15]. After cyanobacterial presence has been confirmed, species specific primers are then used to effectively determine the profile of the freshwater ecosystem being tested. Using the combined microscopic analysis with molecular techniques allows for an effective and efficient method in determining cyanobacterial profiles among freshwater ecosystems. Flow cytometry is another method that could be used as a complement to microscopy and PCR. Flow cytometry can exploit the fact that phytoplankton contain chlorophyll a. A flow cytometer uses a laser and can perform cell differentiation and quantification based on physical characteristics of cells [16]. With the use of these three methods, a successful profile can be generated observing common species at particular water bodies.

1.5. Water chemistry and lake turnover

pH is a water chemistry parameter that is influenced as much by the external environment than it is internal environment of the water body. The pH of water is partially affected by the CO_2 system components (CO_2, H_2CO_3, and CO_3^{2-}). Under basic conditions (pH>7.0), the CO_2 concentration is related to photosynthesis [17]. Since most algal cells (cyanobacteria or phytoplankton) take in CO_2 during their growth process, the pH of the water body falls within a favorable range for growth of a particular genus or species. Some species, at conditions in which the pH is more than 8.6, may be limited in CO_2 uptake due to inactive ion transport mechanisms. But, it is also known that photosynthesis can occur at a pH of 9-10 in some species. Reduction of photosynthesis is noted at pH above 10 in all species [1].

Dissolved oxygen is another water chemistry parameter that is affected by both internal and external environments. As algal blooms grow, eventually they will exhaust all essential nutrients available in the water body. When this occurs, there is a decrease in biomass presence, which eventually leads to decaying of the algal bloom, producing a scum that decreases the underlying water's oxygen. This depletion of dissolved oxygen can lead to several changes that include hypoxia, in which the dissolved oxygen concentration has dropped below 4 mg/L, or anoxia, in which there is no detectable oxygen levels in the water, leading to death among most finfish and shell fish [18].

Seasonal pond or lake turnover could have had a profound effect on the abundance and population shift of phytoplankton when comparing the summer and fall collections. Lake turnover is a natural event that results in the mixing of pond and lake waters, caused by the changing temperatures in surface waters during the shifting of seasons [19, 20]. The density and weight of water change when temperature changes in the freshwater lakes. Water is most dense at 4°C; it becomes less dense when the temperature drops below 4°C, thus rising to the top [20]. This is how fish and other aquatic life can survive during the winter at the floor of the water body, with the warmer water surrounding them towards the sediment [19]. This feature, along with the fact that colder water having a higher capacity for dissolved oxygen, can support the fact that phytoplankton numbers are significantly reduced during the colder months. Pond or lake turnover could affect the phytoplankton survival by keeping the colder water at the surface of the water body, where phytoplanktons need to remain for sunlight and photosynthesis. Because colder surface temperatures do not support phytoplankton growth, the phytoplankton cell numbers and algal blooms should be greatly reduced after the fall turnover occurs, and may return after the spring turnover is complete.

1.6. Algal biomass dynamics in Northern New Jersey freshwater bodies

The New Jersey Department of Environmental Protection (NJDEP) has developed the NJDEP Ambient Lake Monitoring Network, in which lakes and ponds around the state of New Jersey are tested for water quality. The Network tests at least one station and one outlet of each water body. At these stations, the NJDEP tests for total depth, profile depth, Secchi, water temperature, dissolved oxygen, pH, conductivity, phosphorous, nitrates, chlorophyll a, and turbidity, among other water quality factors.

Essex County, New Jersey, is one of the most densely populated counties in the state of New Jersey, consisting of a population of 783,969 in a land area of 127 square miles [21, 22]. Essex County is a heavily urbanized county located in the New York Metropolitan area. Essex County contains 12 major highways, three of the nation's major transportation centers (Newark Liberty International Airport, Port Newark, Penn Station), and 1,673 miles of public roads [21]. These factors, combined with the massive industrial centers producing goods ranging from chemicals to pharmaceuticals, contribute to the urbanization of the area. Despite being heavily urbanized, Essex County has several parks, freshwater rivers, lakes, and ponds which contribute to the continued efforts in beautification and habitat diversity of the region. These bodies of water, being continually subjected to harmful elements from manmade chemicals and excess nutrient pollution, have seen an increase in phytoplankton blooms. Increases in the amounts of nutrients entering lakes and reservoirs in recent decades in urbanized settings as well as associated changes in the water body's biologics have contributed to the increase in focus on the problem of nutrient enrichment due to pollution, called eutrophication [1, 22]. Eutrophication and harmful algal blooms are serious global environmental issues.

In the present study, five lakes in Essex County, New Jersey were sampled in the summer and fall of 2011. Sites were tested for pH, dissolved oxygen, and temperature to observe environmental conditions which harbor algal bloom formation. Samples were subsequently tested for the presence of cyanobacteria and phytoplankton using the three methods described above: Microscopic analysis, polymerase chain reaction, and flow cytometry. Microscopic analysis was performed to identify individual species of cyanobacteria and other phytoplankton among each site at each of the five lakes tested. Once cyanobacteria were confirmed and several species identified, polymerase chain reaction was used with universal primers to confirm the presence of cyanobacteria as well as species specific primers to confirm the presence of particular species. Flow cytometry was utilized to compare seasonal profiles as well as to compare the cyanobacterial cell concentrations among the water samples.

2. Materials and methods

2.1. Cyanobacterial cultures

Synechococcus sp. IU 625 and *Synechococcus elongatus* PCC 7942 strains were used as controls in this study. Five ml of cells were inoculated in 95 ml of sterilized Mauro's Modified Medium (3M) [23] in 250 ml Erlenmeyer flasks [24]. The medium was adjusted to a pH of 7.9 using 1 M NaOH or HCl. The cultures were grown under consistent fluorescent lighting and at a temperature of 27° C. The cultures were grown on an Innova™ 2000 Platform Shaker (New Brunswick Scientific, Enfield, CT, USA) with continuous pulsating at 100 rpm.

2.2. Environmental samples

Water samples were collected from several water bodies in Essex County, New Jersey in 2011. Permission was granted from the Essex County Department of Parks, Recreation and Cultural Affairs for sample collections. There were two collection periods in this study: May 2011-August 2011 (Summer Collections) and October 2011-November 2011 (Fall Collections) to observe microorganism profile seasonal differences. Three to five samples were collected at each body of water, varying in location and water movement. The five bodies of water observed in this study were Diamond Mill Pond (Millburn, NJ, USA), South Orange Duck Pond (South Orange, NJ, USA), Clarks Pond (Bloomfield, NJ, USA), Verona Lake (Verona, NJ, USA), and Branch Brook State Park (Newark, NJ, USA). Before collection, each site was tested for pH, dissolved oxygen, and temperature using the ExStickII® pH/Dissolved Oxygen (DO)/ Temperature meter (ExTech® Instruments corp., Nashua, NH, USA). Samples were collected from each water body in 1 L sterile collection bottles (Nalgene, Rochester, NY, USA). The one liter samples were brought to the lab (Seton Hall University, South Orange, NJ, USA) to be further processed. Each sample was run through a coarse filter with a pore size of 2.7 μm (Denville Scientific,

Metuchen, NJ, USA). Filtrate from the coarse filtered sample was run through a fine fil-ter with a pore size of 0.45 μm (Nalgene, Rochester, NY, USA). Both coarse and fine fil-ter from each sample were placed in a Thelco™ Model 2 incubator for drying at 37°C (Precision Scientific, Chennai, India). Aluminum foil was sterilized by UV light using a Purifier Vertical Clean Bench (Labconco, Kansas City, MO, USA). Dried filters were placed on sterilized aluminum foil and placed in -20°C freezer for further studies.

2.3. Genomic DNA extraction

Genomic DNA of S. IU 625 and S. elongatus PCC 7942 were extracted using Fermentas® Ge-nomic DNA Purification Kit (Fermentas, Glen Burnie, MD, USA). Ten ml of cyanobacteria cells (OD750 nm = ~1) were placed in a 15 ml conical tube. The conical tubes were then cen-trifuged and cells were resuspended in 200 μl of TE Buffer. 200 μl of cells were then mixed with 400 μl lysis solution in an Eppendorf tube (Enfield, CT, USA) and incubated in an Iso-temp125D™ Heat Block (Fisher Scientific, Pittsburgh, PA, USA) at 65°C for 5 minutes. 600 μl of chloroform were added and emulsified by inversion. The sample was then centrifuged at 10,000 rpm for two minutes in a Denville 260D microcentrifuge (Denville Scientific, South Plainfield, NJ, USA). While centrifuging, the precipitation solution was prepared by mixing 720 μl of deionized water with 80 μl of 10X concentrated precipitation solution. After centri-fugation, the upper aqueous phase was transferred to a new tube and 800 μl of the precipita-tion solution were added. The tube was mixed by several inversions at room temperature for two minutes and centrifuged at 10,000 rpm for two minutes. The supernatant was re-moved completely and the DNA pellet was dissolved by adding 100 μl of 1.2 M NaCl solu-tion with gentle vortexing. 300 μl of cold ethanol (100%) was added to enable DNA precipitation and kept in -20°C for 10 minutes. The tube was then centrifuged at 10,000 rpm for three minutes. Ethanol was discarded and the pellet was washed with 70% cold ethanol. The DNA was then dissolved in sterile deionized water, and the DNA concentration and purity were determined with NanoDrop ND-1000 Spectrophotometer (Thermo Fisher Scien-tific, Wilmington, DE, USA).

2.4. Chelex® DNA extraction of environmental samples

All environmental samples underwent a modified Chelex® DNA extraction as follows. Each filter (for both coarse and fine filters) was hole punched 3-4 times at various spots on the filter to produce three to four disks; disks were placed into 1.5 ml Eppendorf tubes. Five hundred microliters of deionized water were added to each tube and each tube was vor-texed. Tubes were let stand for 10-15 minutes. All tubes were centrifuged for three minutes at 10,000 rpm to concentrate the pellet. Clear supernatant was discarded from each tube, and 200 μl of InstaGene Matrix (Bio-Rad Laboratories, Hercules, CA, USA) were added. Each tube was vortexed for 10 seconds. The tubes were incubated for two hours in a Polyscience© Temperature Controller water bath (Polyscience, Niles, IL, USA) at 56° C, vortexed for 10 seconds, and placed in an Isotemp125D™ Heat Block (Fisher Scientific, Pittsburgh, PA, USA) for 8 minutes at 100°C. The tubes were then centrifuged for 10 minutes at 10,000 rpm, and the supernatant (containing DNA) was transferred to clean Eppendorf tubes. The DNA concentration and purity were determined with NanoDrop ND-1000 Spectrophotometer (Thermo Fisher Scientific, Wilmington, DE, USA).

2.5. Polymerase Chain Reaction (PCR)-based assays

DNA extracted from the environmental samples, along with the controls (*S.* IU 625, *S. elongatus* PCC 7942) was amplified using general and specific primers to identify the presence of bacteria, cyanobacteria, phytoplankton, and the dominating species. General primers were used to identify bacteria, cyanobacteria, and phytoplankton by utilizing the bacteria-specific 16s rRNA gene primers 27FB and 785R, PSf and PSr, and CPC1f and CPC1r, respectively. Specific primers were used after phytoplankton and cyanobacteria were detected in the samples. PCR was performed using 6.5 μl nuclease-free deionized water (Promega, Madison, WI, USA), 2.5 μl dimethyl sulfoxide (DMSO), 1 μl of primer in the forward orientation, 1 μl of primer in the reverse orientation, 1.5 μl of DNA sample, and 12.5 μl GoTaq® Hot Start Green Master Mix (Promega). Thermocycling was performed in Veriti 96 Well Thermocycler (Applied Biosystems, Carlsbad, CA, USA). The initial denaturation step was at 95°C for 2 minutes, followed by 35 cycles of DNA denaturation at 95°C for 45 seconds, primer annealing at 50-55°C for 45 seconds, and DNA strand extension at 72°C for 45 seconds, and a final extension step at 72°C for 5 minutes. The amplified DNA was visualized on a 1% agarose gel with ethidium bromide incorporated using TAE electrophoresis buffer (Fermentas). The gel was visualized using a 2UV Transilluminator Gel Docit Imaging System (UVP, Upland, CA, USA).

Primers used in this study were either developed using NCBI BLAST (http://www.ncbi.nlm.nih.gov/BLAST) or by previous studies in this subject field. The sequences of the selected primers, their target organisms and the size of the amplicons are listed in Table 1.

General primers included Phytoplankton-Specific PSf/PSr which identified the 16s rRNA gene in all phytoplankton [17]. Universal primers Uf/Ur identified the 16s rRNA gene in all bacteria [17]. General primers 27FB and 785R were utilized to identify the 16s rRNA in all bacteria, cyanobacteria, and phytoplankton [21]. CPC1f/CPC1r are also cyanobacteria specific primers which identify the β-Subunit of the phycocyanin gene conserved among all cyanobacteria [15]. AN3801f/AN3801r are also cyanobacteria specific primers, identifying the DNA polymerase III gene conserved in *S.* IU625 and *S. elongatus* PCC 7942. Once cyanobacteria and phytoplankton were identified in a sample, specific primers were obtained and utilized. Primers specific for *Anabaena circinalis* toxin biosynthesis gene cluster were developed using NCBI BLAST searches: ANAf and ANAr. Primers to locate Microcystis were developed in accordance with Herry et al. Diatom presence was identified using primers developed in accordance with Baldi et al. 528f with 650r identified the small subunit ribosomal DNA gene conserved among all diatom species [23].

Primer	Target Organism	Sequence	Amplicon (bp)
CPC1f	Cyanobacteria	GGCKGCYTGYYTRCGYGACATGGA	389
CPC1r		AARCGNCCTTGVGWATCDGC	
27fB	Bacteria/ Photosynthetic plankton	AGAGTTTGATCMTGGCTCAG	740
785r		ACTACCRGGGTATCTAATCC	
ANAf	Anabaena circinalis	GATCTAGCCTCACCTGTTGACTT	457
ANAr		GGGATCCTTTTTGCTGCGCC	
528f	Diatoms	GCGGTAATTCCAGCTCCAA	200
650r		AACACTCTAATTTTTTCACAG	
PSf	Phytoplankton	GGGATTAGATACCCCWGTAGTCCT	150
Ur		ACGGYTACCTTGTTACGACTT	
Uf	Phytoplankton	GAGAGTTTGATCCTGGTCAG	700
PSr		CCCTAATCTATGGGGWCATCAGGA	
AN380f	S. IU 625 & S. elongatus PCC 7942	CAAATCACTCAGTTTCTGG	180
AN380r		CAGTAGCAGCTCAGGACTC	

Table 1. Seven primer sets used for PCR-based assays.

2.6. Microscopic analyses

Microscopic images were acquired using a Carl Zeiss AxioLab.A1 phase contrast microscope coupled with a Carl Zeiss AxioCam MRc camera (Carl Zeiss Microimaging, Jena, Germany). Coarse filters (2.7 µm pores) were hole punched one time. The fragment was placed into an Eppendorf tube and 100 µl of deionized water were added. The tubes were left at room temperature for 10-20 minutes. 16 µl of the tube's contents were pipetted onto a microscope slide and viewed at 400X power under the phase filter. Images of diatoms, phytoplankton and cyanobacteria were compared to the atlas "Freshwater Algae of North America: Ecology and Classification" [2]. Species of cyanobacteria, diatoms, and phytoplankton were identified for use in specific PCR analysis and amplification.

2.7. Flow cytometry

Flow cytometry was performed on several sites collected from Branch Brook State Park (Newark, NJ) in June 2011 as well as December 2011 by a Guava® EasyCyte™ Plus Flow Cytometry System (Millipore, Billerica, MA, USA). Fluorescence resulting from the excitation with a 488 nm laser was collected using both green and red filters. A 575 nm filter was used to locate carothenoid pigments, while a 675 nm filter was used to locate chlorophyll a pigments, each of which would be indicative of cyanobacterial presence in the water sample. Blanks were created by using Phosphate Buffered Saline (PBS) and deionized water. Tubes were prepared by hole punching both coarse and find filters and placing them in Eppendorf tubes with 100 μl deionized water, as mentioned above. Cyanobacterial presence was studied in the coarse filter from Branch Brook Park site C from July 2011 (Algal Bloom present), Branch Brook Sites C & D (Raw, unfiltered samples), and both Branch Brook Sites C & D Coarse and Fine filters. Flow Cytometry results were analyzed using FlowJo 7.6.5 Flow Cytometry Analysis Software (Tree Star, Inc., Ashland, OR, USA).

3. Results

3.1. Water chemistry

The pH, dissolved oxygen, and temperature were analyzed at all sites in this study. These parameters aided in the development of a profile for each water body, highlighting which environmental conditions allowed for cyanobacterial and other phytoplankton overgrowth. In Table 1 below, the range of water chemistry levels determined at all sites from summer and fall collection is displayed. The data indicated that the pH range is broader in the fall than in the summer. Dissolved oxygen levels were similar in two seasons and the temperature differences ranged between 6.7 and 20.9 °C.

Water Chemistry	pH	Dissolved Oxygen (mg/L)	Temperature (°C)
Summer	7.27 – 8.30	1-10	23.5 – 30
Fall	6.60-9.25	2-11	9.1-16.8

Table 2. The range of water chemistry parameters for water samples taken at 20 sites during the summer and fall collections is shown.

3.2. Polymerase Chain Reaction (PCR)-based assays

Polymerase chain reaction based assays were performed using DNA extracted from both the coarse and fine filters at each site among all water bodies involved in this study collected in both the summer and fall to identify the presence of bacteria, cyanobacteria, and phytoplankton within each body of water.

PCR-based assays were performed on the coarse and fine filters from each site collected from each water body during the summer and the fall collections. Primer sets used for these PCR-based assays include CPC1f/CPC1r and 27fB/785r for general identification of cyanobacteria and photosynthetic bacteria, respectively. *Synechococcus* sp. IU 625 and *Synechococcus elongatus* PCC 7942, both lab strains, were used as positive controls in this study. In figures 1-3 below, selected gel electrophoresis results from these PCR-based assays are displayed.

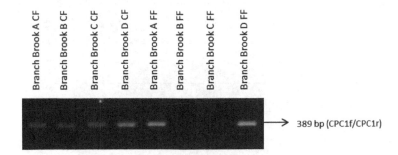

Figure 1. Results from Branch Brook State Park coarse and fine filters are shown using the CPC1f/CPC1r primer set to detect cyanobacteria. It indicates that the presence of cyanobacteria in all 4 sites (A, B, C & D) of Brank Brook State Park.

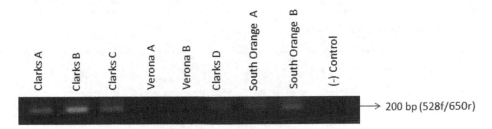

Figure 2. Results from Clarks, Verona and South Orange Duck Pond coarse and fine filters are shown using the 528f/650r primer set to detect diatoms. It indicates that the presence of diatoms in all 4 sites (A, B, C & D) of Clarks Pond and 2 sites of South Orange Duck Pond.

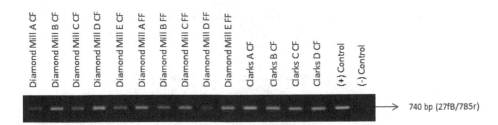

Figure 3. Results from Diamond Mill and Clarks Pond are shown using the 27fB/785r primer set to detect bacteria and photosynthetic phytoplankton. This is indicative of bacterial and photosynthetic phytoplankton presence among all sites tested.

Water bodies	Sites collected	Summer positive detection	Fall positive detection
Brank Brook State Park	A, B, C, D	A, B, C, D	A, B, C, D
Clarks Pond	A, B, C, D, E	A, B, C, D	ND*
Diamond Mill Pond	A, B, C, D	A, C	ND*
South Orange Duck Pond	A, B, C, D	A, B	ND*
Verona Lake	A, B, C	A	A

Table 3. Summary of cyanobacterial detection in the summer and in the fall among 5 water bodies. ND indicates non-detectable. The results showed the fall water samples contain fewer cyanobacteria.

Water bodies	Sites collected	Summer positive detection	Fall positive detection
Branch Brook State Park	A, B, C, D	A, B, C, D	A, B, C, D
Clarks Pond	A, B, C, D, E	A, B, C, D	ND*
Diamond Mill Pond	A, B, C, D	A	ND*
South Orange Duck Pond	A, B, C, D	A, B	ND*
Verona Lake	A, B, C	ND*	ND*

Table 4. Summary of diatom detection in the summer and in the fall among 5 water bodies. ND indicates non-detectable. The results showed the fall water samples contain fewer diatoms.

In summary, PCR-based assays are able to detect cyanobacteria in 65% (13 out of 20) of all the sites collected for summer samples and 25% (5 out of 20) of all sites collected for fall samples. As for diatoms, 55% (11 out of 20) of the sites indicated presence of diatoms while 20% (4 out of 20) of the sites showed positive results. Bacteria and photosynthetic plankton are detected in all sites. This study suggested Branch Brook State Park had the most cyano-bacteria, diatoms and other phytoplankton among 5 water bodies. The result is consistent to the visual algal bloom observed at these sites.

3.3. Microscopic observations

Each coarse and fine filter were hole-punched, re-suspended in De-Ionized water, and observed under a phase contrast microscope in order to detect, verify, and determine abundant species among each water body at each site. Representative images of cyanobacteria and diatoms were displayed in figures 4-7.

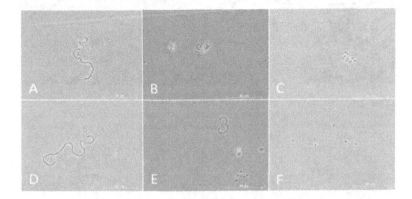

Figure 4. Cyanobacteiral images identified from South Orange Duck Pond. (A) Filamentous cyanobacteria identified from site B. (B) Rod-Shaped cyanobacteria identified from site B. (C) Synechococcus identified from site B. (D) Filamentous cyanobacteria identified from site C. (E) Filamentous cyanobacteria identified from site C. (F) Synechococcus identified from site B. (1000X)

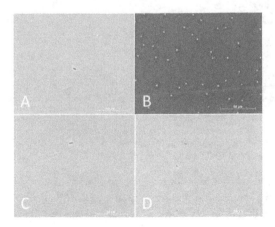

Figure 5. Cyanobacteria identified from Clarks Pond in Bloomfield, NJ. (A) Synechococcus identified from site B. (B) Cyanobacteria identified from site B. (C) Synechococcus identified from site C. (D) Synechococcus identified from site D. (1000X)

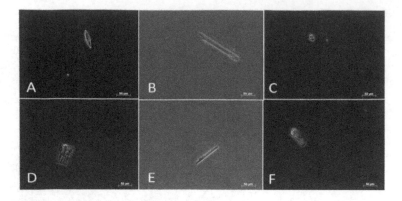

Figure 6. Diatom images identified from Verona Lake in Verona, NJ. (A) Diatom identified from site A. (B) Diatom identified from site B. (C) Diatom identified from site A. (D) Diatom- Fragilaria identified from site A. (E) Diatom identified from site A. (F) Diatom- Fragilaria identified from site B. (400X)

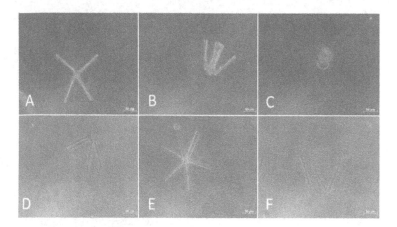

Figure 7. Images of Diatoms at Branch Brook State Park in Newark, NJ. (A) Diatom - Asterionella identified from site A. (B) Diatom - Asterionella identified from site A. (C) Diatom identified from site A. (D) Diatom - Asterionella identified from site B. (E) Diatom - Asterionella identified from site B. (F) Diatom - Asterionella identified from site B. (400X)

A comparison was constructed in order to study the effectiveness of using both PCR and microscopic analysis in identification of the common species of cyanobacteria and other phytoplankton in the water bodies in this study.

Microscopic observation suggested that most microbes among the water sample collected were bacteria, cyanobacteria and diatoms. Cell density were determined and recorded during microscopic analysis from each site of the freshwater ecosystems in this study. Cell density was calculated and was subsequently plotted against water chemistry parameters, including pH, dissolved oxygen, and temperature for each site during the summer and fall collections.

Water Body	Site	Diatoms	Cyanobacteria	Photosynthetic Bacteria
Branch Brook Lake	A	PCR & MI	PCR & MI	PCR & MI
	B	PCR & MI	PCR & MI	PCR & MI
	C	PCR & MI	PCR & MI	PCR & MI
	D	PCR & MI	PCR & MI	PCR & MI
Clarks Pond	A	PCR	PCR	PCR
	B	PCR	PCR	PCR
	C	PCR	PCR	PCR
	D	PCR	PCR	PCR
Diamond Mill Pond	A	PCR & MI	PCR & MI	PCR & MI
	B	PCR & MI	MI	PCR & MI
	C	PCR & MI	PCR & MI	PCR & MI
	D	MI	MI	PCR
	E	PCR & MI	MI	PCR & MI
South Orange Duck Pond	A	PCR	PCR & MI	PCR & MI
	B	PCR	PCR	PCR
	C	MI	MI	PCR
	D	MI	MI	PCR
Verona Lake	A	PCR & MI	PCR & MI	PCR & MI
	B	PCR & MI	MI	PCR & MI
	C	PCR & MI	MI	PCR & MI

Table 5. The correlation between microscope findings and PCR findings from summer collections is depicted.

Water Body	Site	Diatoms	Cyanobacteria	Photosynthetic Bacteria
Branch Brook Lake	A	PCR & MI	PCR & MI	PCR & MI
	B	PCR & MI	PCR	PCR & MI
	C	PCR	PCR	PCR & MI
	D	PCR	PCR	PCR & MI
Clarks Pond	A	MI	MI	PCR & MI
	B	MI	MI	PCR & MI
	C	MI	MI	PCR & MI
	D	MI	MI	PCR & MI
Diamond Mill Pond	A	MI	MI	PCR & MI
	B	MI	MI	PCR & MI
	C	MI	MI	PCR & MI
	D	MI	MI	PCR & MI
	E	MI	MI	PCR & MI
South Orange Duck Pond	A	MI	MI	PCR & MI
	B	MI	MI	PCR & MI
	C	MI	MI	PCR & MI
	D	MI	MI	PCR & MI
Verona Lake	A	MI	PCR	PCR & MI
	B	MI	MI	PCR & MI
	C	MI	MI	PCR & MI

Table 6. The correlation between microscope findings and PCR findings from fall collections is depicted.

4. Discussion

Each water chemistry factor (pH, DO, temperature) was determined at each site of each water body to gain insight on the environmental conditions that harbor eutrophication and/or algal bloom production. A combination of these physical and chemical properties along with biotic features of the natural water bodies work in a symbiotic manner to determine the sensitivities of these water bodies to eutrophication [18].

For the summer collections, the pH ranged between 7.27 and 9.20 among all 20 sites. The pH and cell density agree with the fact that there is a certain pH range which favors growth. There are several sites in which the pH was found to be between 7 and 8.5, in which both the highest density of cyanobacteria and diatom were found. As the pH drops below 7, however, there were no visible cyanobacteria or diatoms. As the pH increases to over 8.5, the cell density and the amount of visible cells in the samples visibly decrease. It is understood that the optimal pH range for cyanobacteria growth is found to be between 7.5 and 10 [26]. During fall collections, the pH ranged from 6.60 to 9.25, which again reveals an alkaline environment except for one site (Branch Brook site C). The sites with pH ranging between 7 and 8.5 appear to contain the highest cell count of both diatom and cyanobacteria.

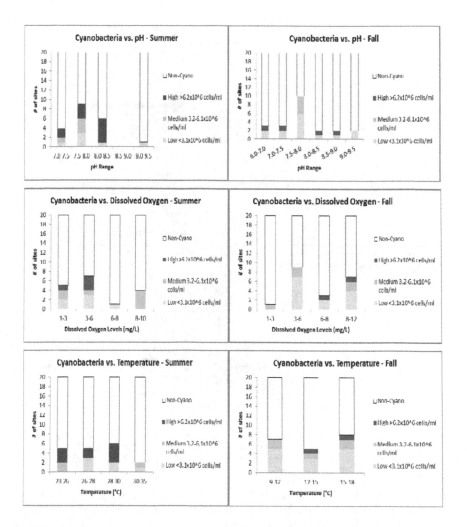

Figure 8. A comparison between cyanobacterial cell density from the summer and the fall collections. Water chemistry parameters include pH, dissolved oxygen, and temperature. the graph shows the number of sites with a high cyanobacteria cell density (>6.2x10⁶cells/ml) (Red), sites with a medium cyanobacteria cell density (3.2x10⁶ − 6.0x10⁶ cells/ml) (Green), and sites with a low cyanobacteria cell density (<3.1x10⁶cells/ml) (Grey).

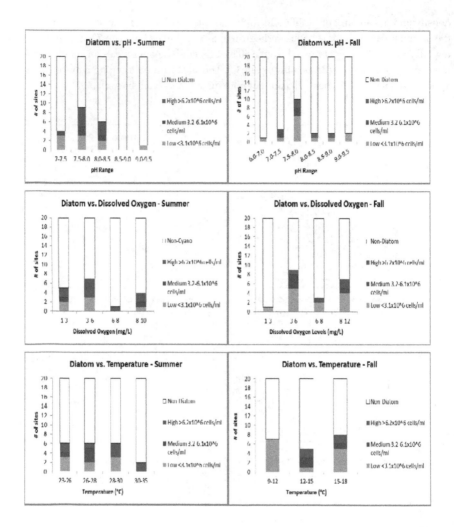

Figure 9. A comparison between diatom cell density from the summer and the fall collections. Water chemistry parameters include pH, dissolved oxygen, and temperature. the graph shows the number of sites with a high diatom cell density (>6.2x10⁶cells/ml) (Red), sites with a medium diatom cell density (3.2x10⁶ – 6.0x10⁶ cells/ml) (Green), and sites with a low diatom cell density (<3.1x10⁶cells/ml) (Grey).

Among the ponds and lakes tested in the summer collections, dissolved oxygen levels ranged from 1 mg/L to 10 mg/L, showing a wide range of dissolved oxygen levels. Because it has been previously reported that algal blooms are known to decrease the dissolved oxygen levels [18, 27], it was important to detect a profile of cells found at each dissolved oxygen level. In figures 34 and 35 above, the graph shows both more cyanobacteria as well as diatom cell numbers recorded at sites with a lower dissolved oxygen level (<5 mg/L). This result is important because it displays the correlation between cyanobacteria and other phytoplankton growth and the significant depletion of oxygen in the water column. During the fall collections, the dissolved oxygen levels ranged from 2 mg/L to 11 mg/L, representing a slight increase in dissolved oxygen which may be a result of decrease in biomass among the water bodies tested, or as a result of lake turnover. Although there are equal numbers among the four sites with countable cell numbers, the cell density are minute when compared to the summer collections. The diatom cell densities appear to be higher among lower dissolved oxygen, although there are still some sites in the higher dissolved oxygen range (8-10 mg/L) that contain cyanobacteria.

The last factor of water chemistry incorporated into this study was temperature of each site. Temperature has not only been shown to affect the cell size of phytoplankton by controlling enzymatic reactions within the cells, but also to regulate the multiplication rate and standing biomass (phytoplankton population) within the water body. Studies have also shown, though, that it may not be temperature that is limiting the fall and winter growth but the lack of sunlight for photosynthesis [1, 28]. Water temperature seems to dictate the phytoplankton profile. For example, the cyanobacteria Anabaena has been found to be severely affected by lower temperatures while the diatom Asterionella is not as affected by temperature but a decrease of nutrients in the water body. During the summer collections, the temperatures ranged from 23.5 to 30.2°C. The amount of cyanobacteria and diatom cells appears to be at a peak between 25 and 30 °C. This is important because it has been previously reported that the optimal growth rate of 'algae' (phytoplankton cells) is between 20 and 25°C [1]. As temperature increased from the lowest recorded (23.5 °C), there was a clear increase in both cyanobacteria and diatom cell number, which seemed to decrease after 30°C. Also, when comparing the phytoplankton distribution between summer collections and fall collections, there is a clear separation in the cell count between the two seasons. This is another finding that corresponds with previous studies that the combination of decreased temperature and decreased light availability for photosynthesis results in decreased phytoplankton growth rate [1, 29]. The temperature of sites during the fall fell between the ranges of 9.1 and 17.9°C. As stated above and seen in figures 9 and 10, the cell count of both diatom and cyanobacteria cells are greatly reduced. The higher cell counts of both cyanobacteria and diatoms in the fall collections appeared to fall in the temperature range of between 13 and 17.9°C. Below 13°C, there were no readily countable cyanobacteria or diatom cells. This could be due to limited cell growth below the optimal growth rate temperature. Between 13 and 17.9°C, there were cyanobacteria and diatoms, although at a clearly decreased level when compared to the summer collections and observations. Lake or pond turnover, with a combination of decreasing temperature and increasing dissolved oxygen levels, may have resulted in a decreased amount of phytoplankton cells between summer and fall collections.

Polymerase Chain Reaction (PCR) provided additional verification on the presence of bacteria, cyanobacteria, and algae in the freshwater ecosystems observed. In order to identify phytoplankton, PCR was employed. In order to identify these sequences, housekeeping gene sequences were exploited. Housekeeping genes are constantly present and active within living cells, but they do not have to be activated to be identified. The small-subunit ribosomal RNA (16S rRNA) gene segment is a housekeeping gene found within all phototrophs. Because the 16S rRNA gene segment is always present within the genome of cyanobacteria and bacteria cells, this gene segment served as the target for environmental PCR. The universal primers used in this study (27fB/785r, PSf/Ur, Uf/PSr) utilized the 16s rRNA segment to identify cyanobacteria, bacteria, and phytoplankton have been identified in previous studies [14, 15, 30]. The specific primer used in this study to identify the species *Anabaena circinalis* was developed using NCBI BLAST, while primers for diatoms and Microcystis have been identified in previous studies [31, 32]. In both the summer and fall collections, PCR proved successful in detecting the presence of cyanobacteria, bacteria and phytoplankton by the general "universal" primers identified in previous studies. After detection of these cells within the lakes, a general profile was constructed for each lake.

Branch Brook State Park site C developed a clearly visible algal bloom during the summer collections in July, 2011. During microscopic observation of the coarse filter collected from this site, several species of cyanobacteria were detected. A species of Oscillatoria was detected at site C. Oscillatoria is a type of filamentous cyanobacteria. Oscillatoria, along with other filamentous cyanobacteria, has been previously reported to cause algal blooms [34]. Another cyanobacterium, Radiococcus, was identified at Branch Brook State Park site C. Species of Radiococcus have been detected in small numbers in previous studies [35], but it has not been recorded to cause algal blooms. Because Radiococcus was seen at increased numbers at this site, this cyanobacterium may have been another factor in this algal bloom production and persistence.

Tables 5 and 6 above show the relationship found between microscopy and PCR. Both tables indicate the similarities found between observations made under microscopic observation and PCR. These results prove that although PCR and microscopy may be inefficient on their own, together they are an effective mechanism to develop a phytoplankton profile for freshwater lakes. These findings correlate with previous studies, which have found that it is difficult to distinguish similar cell morphologies by microscopy [36]. Although it is inefficient, microscopy still remains the preeminent means for morphotyping, cell counting, biovolume, viability assays, and life cycle stage observations of cells in a cyanobacterial or phytoplankton bloom [36]. The combination of the microscopic technique and the molecular technique provided for a well detailed and wide analysis of the five ecosystems tested in this study.

Flow cytometry provided a rapid analysis for the overall profile of the sites being tested. Flow cytometry was used to detect the overall photoautotroph presence in the sample by exploiting the auto fluorescence mechanisms of all cells containing the photosynthetic pigment chlorophyll *a*. Flow cytometry was able to show the amount of phycocyanin-containing cells when compared to total cells in the sample.

5. Conclusion

Modified Chelex® DNA extraction is an efficient way to isolate DNA from environmental water samples. When it comes to species identification, PCR-based assays appeared to be more rapid and sensitive than microscopic observation on cyanobacteria and other phytoplankton. Microscopic observation aided in identification of the common genera, while PCR was allowed for identification up to the species level. In addition, flow cytometry was able to provide insight on the phytoplankton profile when used in conjunction with the other two methods. The combination of the three methods can be employed to provide a thorough analysis of the water bodies observed in the study Microscopic observation also allowed for cell density determination, which was important in seasonal comparisons. Water chemistry parameters (pH, DO, and temperature) were crucial to be incorporated in order to establish the correlation between phytoplankton profile and environmental conditions.

6. Future studies

In order to obtain a larger, more complete profile of phytoplankton growth in New Jersey freshwater ecosystems, flow cytometry must be employed at a larger scale. In the current study, it has been established that flow cytometry is successful in the detection of cells containing chlorophyll a. In order to develop a rapid yet large profile for many freshwater ecosystems, fluorescent probes must be employed. As used in PCR, the 16s rRNA segments found in all phototrophs can be detected and probed with fluorescence. With these probes, the flow cytometer can successfully identify different species of cyanobacteria and phytoplankton while analyzing mixed microbial populations [25].

As mentioned above, phosphates and nitrates are two of the most important elements resulting from pollution that drive the eutrophication in freshwater ecosystems. Also, as the biomass of cyanobacteria and phytoplankton increase, the amount of chlorophyll a will increase. These factors are important in monitoring and identifying ecosystems that are threatening for algal bloom formation. In order to gain a complete profile for each freshwater ecosystem, these parameters must be incorporated into the future study.

Author details

Tin-Chun Chu and Matthew J. Rienzo

Department of Biological Sciences, Seton Hall University, South Orange, NJ, USA

References

[1] C.S. Reynolds, The Ecology of Freshwater Phytoplankton, Cambridge University Press, 1984.

[2] J.D.Wehr, R.G. Sheath, P. Kociolek, and J.H. Thorp, Freshwater Algae of North America: Ecology and Classification (Aquatic Ecology), Elsevier Science, 2002.

[3] M.J. Leng, and P.A. Barker, "A review of the oxygen isotope composition of lacustrine daitom silica for paleoclimate reconstruction," Earth-Science Reviews, 2005, p. 5-27.

[4] B.J.F. Biggs, and C. Kilroy, Stream Periphyton Monitoring Manual, The Crown (acting through the Minister for the Environment), 2000.

[5] A. Zingone, and H.O. Enevoldsen, "The diversity of harmful algal blooms: a challenge for science and management," Ocean & Coastal Management, 2000, p. 725-48.

[6] S.B. Watson, E. McCauley, and J.A. Downing, "Patterns in phytoplankton taxonomic composition across temperate lakes of differing nutrient status," ASLO: Limnology and Oceanography, 1997, p. 487-95.

[7] C.R. Anderson, M.R.P. Sapiano, M.B.K. Prasad, W. Long, P.J. Tango, C.W. Brown, and R. Murtugudde, "Predicting potentially toxigenic Pseudo-nitzschia blooms in the Chesapeake Bay," Journal of Marine Systems, 2010, p. 127-40.

[8] T.C. Chu, S.R. Murray, S.F. Hsu, Q. Vega, and L.H. Lee, "Temperature-induced activation of freshwater Cyanophage AS-1 prophage," Acta Histochemica, May 2011, p. 294-9.

[9] G.A. Codd, L.F. Morrison, and J.S. Metcalf, "Cyanobacterial toxins: risk management for health protection," Toxicology and Applied Pharmacology, 2005, p. 264-72.

[10] EPA, "Water: Contaminant Candidate List (CCL 2)," 2005.

[11] J. Al-Tebrineh, T.K. Mihali, F. Pomati, and B.A. Neilan, "Detection of saxitoxin-producing cyanobacteria and Anabaena circinalis in environmental water blooms by quantitative PCR," Applied and Environmental Microbiology, 2010, p. 7836-42.

[12] CDC, "Facts about Cyanobacteria & Cyanobacterial Harmful Algal Blooms," 2009.

[13] L.H. Lee, D. Lui, P.J. Platner, S.F. Hsu, T.C. Chu, J.J. Gaynor, Q.C. Vega, and B.K. Lustigman, "Induction of temperate cyanophage AS-1 by heavy metal – copper," BMC Microbiology, 2006, 6:17.

[14] U. Nübel, F. Garcia-Pichel, and G. Muyzer, "PCR primers to amplify 16S rRNA genes from cyanobacteria," Applied and Environmental Microbiology, 1997, p. 3327-32.

[15] J.W. Stiller, and A. McClanahan, "Phyto-specific 16S rDNA PCR primers for recovering algal and plant sequences from mixed samples," Molecular Ecology Notes, 2005, p. 1-3.

[16] F. Palumbo, G. Ziglio, and A. Van der Beken, Detection methods for algae, protozoa and helminths in fresh and drinking water, John Wiley and Sons, 2002.

[17] F. Moatar, F. Fessant, and A. Poirel, "pH modelling by neural networks. Application of control and validation data series in the Middle Loire river," Ecological Modelling, 1999, p. 141-56.

[18] H.W. Paerl, R.S. Fulton III, P.H. Moisander, and J. Dyble, "Harmful freshwater algal blooms, with an emphasis on cyanobacteria," The Scientific World Journal, 2001, p. 76-113.

[19] MDOC, "Aquaguide: Pond Turnover," Missouri Department of Conservation, 2010, p. 4021_2870.

[20] D.N. Castendyk, and J.G. Webster-Brown, "Sensitivity analyses in pit lake prediction, Martha Mine, New Zealand 1: Relationship between turnover and input water density," Chemical Geology, 2007, p. 42-55.

[21] N.J. Essex County, The County of Essex, New Jersey, 2012.

[22] D.M. Anderson, "Approaches to monitoring, control and management of harmful algal blooms (HABs)," Ocean & Coastal Management, 2009, p. 342-7.

[23] W.A. Kratz, and J. Myers, "Nutrition and Growth of Several Blue-Green Algae," American Journal of Botany, 1955, p. 282-7.

[24] T.C. Chu, S.R. Murray, J. Todd, W. Perez, J.R. Yarborough, C. Okafor, and L.H. Lee, "Adaption of Synechococcus sp. IU 625 to growth in the presence of mercuric chloride," Acta Histochemica, 2012, p. 6-11.

[25] R.I. Amann, B.J. Binder, R.J. Olson, S.W. Chisholm, R. Devereux, and D.A. Stahl, "Combination of 16s rRNA-targeted oligonucleotide probes with flow cytometry for analyzing mixed microbial populations," Applied and Environmental Microbiology, 1990, p. 1919-25.

[26] N. Giraldez-Ruiz, P. Mateo, I. Bonilla, and F. Fernandez-Piñas, "The relationship between intracellular pH, growth characteristics and calcium in the cyanobacterium Anabaena sp. strain PCC7120 exposed to low pH," New Phytologist, 1997, p. 599-605.

[27] M.L. Saker, M. Vale, D. Kramer, and V.M. Vasconcelos, "Molecular techniques for the early warning of toxic cyanobacteria blooms in freshwater lakes and rivers," Applied Microbial Biotechnology, 2007, p. 441-9.

[28] M.G. Alam, N. Jahan, L. Thalib, B. Wei, and T. Maekawa, "Effects of environmental factors on the seasonally change of phytoplankton populations in a closed fresh-water pond," Environment International, 2001, p. 363-71.

[29] R.W. Sterner, and J.P. Grover, "Algal growth in warm temperature reservoirs; kinetic examination of nitrogen, temperature, light, and other nutrients," Water Research, 1998, p. 3539-48.

[30] A.L. Barkovskii, and H. Fukui, "A simple method for differential isolation of freely dispersed and partical-associated peat microorganisms," Journal of Microbiological Methods, 2004, p. 93-105.

[31] F. Baldi, C. Facca, D. Marchetto, T.N. Nguyen, and R. Spurio, "Diatom quantification and their distribution with salinity brines in costal sediments of Terra Nova Bay (Antarctica)," Marine Environmental Research, 2011, p. 304-11.

[32] E. Herry S., A. Fathalli, A.J. Rejeb, and N. Bouaïcha, "Seasonal occurrence and toxicity of Microcystis spp. and Oscillatoria tenuis in the Lebna Dam, Tunisia," Water Research, 2008, p. 1263-73.

[33] D.M. Anderson, P.M. Glibert, and J.M. Burkholder, "Harmful Algal Blooms and Eutrophication: Nutrient Sources, Composition, and Consequences," Estuaries, 2002, p. 704-26.

[34] R.J. Montealegre, J. Verreth, K. Steenbergen, J. Moed, and M. Machiels, "A dynamic simulation model for the blooming of Oscillatoria agardhii in a monomictic lake," Ecological Modelling, 1995, p. 17-24.

[35] C.E. Taft, and W.J. Kishler, Algae from Western Lake Erie, The Ohio State University, The Ohio Journal of Science, 1968, p.80-3.

[36] K.A. Kormas, S. Gkelis, E. Vardaka, and M. Moustaka-Gouni, "Morphological and molecular analysis of bloom-forming Cyanobacteria in two eutrophic, shallow Mediterranean lakes," Limnologica-Ecology and Management of Inland Waters, 2011, p. 167-73.

Chapter 3

The Performance Evaluation of Anaerobic Methods for Palm Oil Mill Effluent (POME) Treatment: A Review

N.H. Abdurahman, Y.M. Rosli and N.H. Azhari

Additional information is available at the end of the chapter

1. Introduction

Palm oil mill effluent (POME) is an important source of inland water pollution when released into local rivers or lakes without treatment. In the process of palm oil milling, POME is generated through sterilization of fresh oil palm fruit bunches, clarification of palm oil and effluent from hydro-cyclone operations [Borja et al.,1996a]. POME is a viscous brown liquid with fine suspended solids at pH ranging between 4 and 5 [Najafpour et al., 2006]. In general appearance, palm oil mill effluent (POME) is a yellowish acidic wastewater with fairly high polluting properties, with average of 25,000 mg/l biochemical oxygen demand (BOD), 55,250 mg/l chemical oxygen demand (COD) and 19,610 mg/l suspended solid (SS). This highly polluting wastewater can cause several pollution problems. Therefore, direct discharge of POME into the environment is not encouraged due to the high values of COD, BOD and SS.

Over the past 20 years, the technique available for the treatment of POME in Malaysia has been biological treatment, consisting of anaerobic, facultative and aerobic pond systems [Chooi, 1984], and [N. Ma, 1999]. Anaerobic digestion has been employed by most palm oil mills as their primary treatment of POME [Tay, 1991]. More than 85% of palm oil mills in Malaysia have adopted the ponding system for POME treatment [Ma et al., 1993], while the rest opted for open digesting tanks [Yacop et al., 2005]. These methods are regarded as a conventional POME treatment method involving long retention times and large treatment areas. High-rate anaerobic bioreactors have also been applied in laboratory-scaled POME treatment such as up-flow anaerobic sludge blanket (UASB) reactors [Borja el al., 1994a]; up-flow anaerobic filtration [Borja et al., 1994b]; fluidized bed reactors [Borja et al., 1995a], [Borja et al., 1995b] and up-flow anaerobic sludge fixed-film (UASFF) reactors [Najafpour et al., 2006]. Anaerobic contact digesters Ibrahim et al. (1984) and continuous stirred tank reactors (CSTR) have also been studied for PMOE treatment Chin (1981). Other than anaerobic digestion, POME has also been treated using membrane technology [Ahmad et al., 2006; 2007] and [Fakhru'l-Razi, 1994].

2. Anaerobic digestion

Anaerobic digestion is the most suitable method for the treatment of effluents containing high concentration of organic carbon such as POME [Borja et al.,1996a]. Anaerobic digestion is defined as the engineered methanogenic anaerobic decomposition of organic matter. It involves different species of anaerobic microorganisms that degrade organic matter [Cote et al., 2006]. In the anaerobic process, the decomposition of organic and inorganic substrate is carried out in the absence of molecular oxygen. The biological conversion of the organic substrate occurs in the mixtures of primary settled and biological sludge under anaerobic condition followed by hydrolysis, acidogenesis and methanogenesis to convert the intermediate compounds into simpler end products as methane (CH4) and carbon dioxide (CO2) [Gee et al., 1994], [Guerrero et al., 1999], and [Gerardi, 2003]. Therefore, the anaerobic digestion process offers great potential for rapid disintegration of organic matter to produce biogas that can be used to generate electricity and save fossil energy [linke, 2006]. The suggested anaerobic treatment processes for POME include anaerobic suspended growth processes, attached growth anaerobic processes (immobilized cell bioreactors, anaerobic fluidized bed reactors and anaerobic filters), anaerobic blanket processes (up-flow anaerobic sludge blanket reactors and anaerobic baffled reactors), membrane separation anaerobic treatment processes and hybrid anaerobic treatment processes.

2.1. Anaerobic and alternative POME treatment methods

Currently available alternative methods for POME treatment are: aerobic treatment, membrane treatment systems and the evaporation method. The advantages and disadvantages of anaerobic and alternative treatment methods are shown in Table 1. In terms of energy requirement for POME treatment operation, anaerobic digestion has a greater advantage over the other alternative methods as it does not require energy for aeration. Furthermore, anaerobic POME treatment produces methane gas (CH4) which is a value-added product to digestion that can be utilized in the mill to gain more revenue in terms of CER. For example, the open digesting tank for POME treatment without land application, the capital cost quoted by [Gopal et al., 1986] for a palm oil mill processing 30 tons FFB/h is RM 750,000. Based on the chemical Engineering Plant Cost Index [Ulrich et al., 2004] the capital cost for this system is estimated to be US 370,272 in 2006. Comparing this to the capital cost for a membrane system in POME treatment for a palm oil mill processing 36 tons FFB/h at RM 3,950,000 [Chong, 2007], it is obvious that the former anaerobic treatment has better advantage over other treatment methods in terms of capital cost. The disadvantages of anaerobic treatment are (a) long retention times and (b) long start-up period. However, the problem of long retention times can be rectified by using high-rate anaerobic bioreactors while the long start-up period can be shortened by using granulated seed sludge [McHugh et al., 2003], utilizing seed sludge from same process [Yacob et al., 2006b] or maintaining suitable ph and temperature in the high-rate anaerobic bioreactor for growth of bacteria consortia [Liu et al., 2002].

Treatment types	Advantages	Disadvantages	Reference
Membrane	Produce consistent and good water quality after treatment, smaller space required for membrane treatment plants, can disinfect treated water	Short membrane life, membrane fouling, expensive compared to conventional treatment	[Ahmad et al., 2006] [Metcalf et al., 2003]
Anaerobic	low energy requirements (no aeration),Producing methane gas as a valuable end product, generated sludge from process could be used for land applications	Long retention time, large area required For conventional digesters, slow start-up (granulating reactors)	[Metcalf et al., 2003] [Borja et al., 2006a]
Evaporation	solid concentrate from process can be utilized as feed material for fertilizer manufacturing	High energy consumption	[MA et al., 1997]
Aerobic	Shorter retention time, more effective in handling toxic wastes	High energy requirement (aeration), rate of pathogen inactivation is lower in aerobic sludge compared to anaerobic sludge, thus unsuitable for land applications	[Jr et al., 1999] [Doble et al., 2005]

Table 1. Advantages and disadvantages between anaerobic and alternative treatment methods

2.2. Anaerobic treatment methods

2.2.1. Anaerobic filtration

Anaerobic digestion has existed as a technology over 100 years. It gradually evolved, from an airtight vessel and a septic tank, to a temperature controlled, completely mixed digester, and finally to a high rate reactor, containing a density of highly active biomass. The microbiology of methane digestion has been examined intensively in the last decade. It has been established that three physiological groups of bacteria, converting hydrogen and carbon dioxide or acetate to methane. In contrast to aerobic degradation, which is mainly a single species phenomenon, anaerobic degradation proceeds as a chain process, in which several sequent organisms are involved. Anaerobic conversion of complex substrates requires the synergistic action of the micro-organisms involved. A factor of utmost importance, in the overall process, is the partial pressure of hydrogen and the thermodynamics linked to it. This fact has been recognized and discussed by researchers [Bryant et al., 1967]; [Boone and Bryant, 1980]; [McInerney et al. 1979]; [Hickey and Switzenbaum 1988]. Anaerobic filter were favor for wastewater treatment because (a) high substrate removal efficiency (b) it requires a smaller reactor volume which operates on a shorter hydraulic retention times (HRT), [Borja et al., 1994b], (c) the ability to maintain

high concentration of biomass in contact with the wastewater without affecting treatment efficiency [Reyes et al., 1999], [Wang et al. (2007)], and (d) tolerance to shock loadings [Reyes et al.,
1999], [Van Der Merwe et al., 1993]. Besides, construction and operation of anaerobic filter is
less expensive and small amount of suspended solids in the effluent eliminates the need of solid separation or recycle [Russo et al. 1985].

Another factor of fundamental importance has been the identification of new methanogenic
species, and the characterization of their physiological behaviour. Of particular interest was
the determination of the substrate affinity constants of both hydrogenotrophic and acetotro-
phic methanogens. While the first exhibit quite high substrate affinities and remove hydrogen down to ppm levels, the second group appears to contain species with only low
substrate affinities [Zehnder et al., 1980]; [Huser et al., 1982]. This limited substrate affinity
has,, an important consequence for anaerobic wastewater treatment.

A technological advance of utmost importance in anaerobic digestion has been the development of methods to concentrate methanogenic biomass in the reactor, especially in very low
solids concentration in the wastewater, 1 - 2%. Such higher concentration of biomass can be
achieved using of autoflocculation and gravity settling as, for instance, in the UASB reactor
[Lettinga et al. 1983], by attachment to a static carrier (anaerobic filter) [Henze and Harre-
moes, 1982]; [Van Den Berg and Kennedy 1981]; [Young and McCarty 1969], by attachment
to a mobile carrier (fluidized bed) [Binot et Heijnen 1984]; [Bull et al., 1984] or by growth in
and on a matrix [Huysman et al., 1983]. All these different methods are in full development

Anaerobic filters have been applied to treat various types of wastewater including soybean
processing wastewater [H-Q et al., 2002a], wine vinases [Nebot et al., 1995], [Perez et al.,
1998], land fill leachate [Wang et al., 2007], municipal wastewater [Bodkhe, 2008], brewery
wastewater [Leal et al., 1998], slaughterhouse wastewater [Ruiz et al., 1997], drug wastewater [Gangagni et al., 2005], and beet sugar water [Farhadian et al., 2007]. However, filter
clogging is a major drawback in the continuous operation of anaerobic filters [Bodkhe,
2008], [Jawed et al. 2000], [Parawira et al., 2006]. Clogging of anaerobic filter has only been
reported in the treatment of POME at an organic loading rate (OLR) of 20 g COD/l/day [Bor-
ja et al., 1995b] and also in the treatment of slaughterhouse wastewater at 6 g COD/l/day.
This because the other studies were conducted at lower OLRs which had lower suspended
solid content compared to POME. In general, anaerobic filter s are capable of treating wastewaters to obtain good effluent quality with at least 70% of COD removal efficiency with
methane gas composition of more than 50%. Table 2 illustrates the COD removal efficiency
of some treated wastewater using anaerobic filtration based on highest achievable percentage of methane in the generated biogas. In terms of POME treatment, the highest COD removal efficiency recorded was 94% with 63% of methane at an OLR of 4.5 kg COD/m3/day,
while overall COD removal efficiency was up to 90% with an average methane gas composition of 60% [Borja et al., 1994b]. Investigations have been done to improve the efficiency of
anaerobic filtration in wastewater treatment. [Yu et al., 2002a] found that operating at an optimal recycle ratio which varies depending on OLR will enhance COD removal. However,
methane percentage will be compromised with increase in optimal recycle ratio. Higher retention of biomass in the filter will also lead to a better COD removal efficiency.

Types of Wastewater	Operating OLR range (Kg COD/m³/day)	COD removal efficiency (%)	Highest methane composition (%)	Reference
Slaughterhouse wastewater	1.0-6.5	79.9 (91.5)	51.1	[Ruiz et al., 1997]
POME	1.2-11.4	94.0 (94.0)	63.0	[Borja et al. 1994b]
Baker's yeast factory effluent	1.8-10.0	69.0 (74.0)	65.0	[Van der et al. 1993]
Distillery wastewaters	0.42-3.4	91.0 (93.0)	63.0	[Russo et al., 1985]
Landfill leachate	0.76-7.63	90.8 (90.8)	N/A	[Wang et al., 2007]

() - number in bracket denotes highest COD removal efficiency. N/A- data unavailable.

Table 2. Operating OLR range; COD removal efficiency in various wastewater treatments using anaerobic filtration based on highest % of methane production

2.2.2. Fluidized bed reactor

A fluidized bed reactor (FBR) is a type of reactor device that can be used to carry out a variety of multiphase chemical reactions. Fluidized bed reactor exhibits several advantages that make it useful for treatment of high-strength wastewaters. It has very large surface areas for biomass attachment [Borja et al., 2001], [Toldra et al., 1987] enabling high OLR and short HRTs during operation [Garcia et al., 1998], [Sowmeyan et al., 2008]. Furthermore, fluidized bed has minimal problems of channeling, plugging or gas hold-up [Borja et al., 2001], [Toldra et al. 1987]. Higher up-flow velocity of raw POME is maintained for fluidized bed reactor to enable expansion of the support material bed. Biomass will then attach and grow on the support on material. In this way, biomass can be retained in the reactor. Hickey and [Switzenbaum, 1988] reported on the development of the anaerobic expanded bed process, which was found to convert dilute organic wastes to methane at low temperatures and at high organic and hydraulic loading rates. This process was being evaluated in 1988, on a 10,000 gallons per day pilot scale, consisting of an anaerobic expanded bed followed by post- treatment. [Jeris, 1987] reported on a two year experiment, testing two pilot scale anaerobic fluidized bed reactors, treating primary effluent. One reactor used sand as a carrier, the other granular activated carbon (GAC). Seeding experiments indicated that the GAC developed a biofilm more quickly and had more attached biomass. In addition, better BOD removal was observed with the GAC reactor. He noted that removal efficiencies were essentially independent of organic volumetric loading rates. Over a twelve month period in temperate climates, effluent total BOD5 values were consistently around 40 mg/l.

Investigations have been done on the application of fluidized bed to treat cutting-oil wastewater [Perez et al., 2007]; real textile wastewater [Sen et al., 2003]; slaughterhouse wastewater [Toldra et al., 1987]; wine and distillery wastewater [Garcia et al. 1998], [Sowmeyan et al.,

2008]; ice-cream wastewater [Borja et al., 1995a], [Hawkes et al., 1995]; pharmaceutical effluent [Saravanane et al., 2001], and POME [Borja et al., 1995b]. OLR ranges and COD removal efficiencies of various wastewater treatments using fluidized bed is tabulated in Table 3. Based on Table 3, it can be concluded that anaerobic fluidized bed can typically remove at least 65% and up to more than 90% of COD. Inverse flow anaerobic fluidized bed is capable of tolerating higher OLRs compared to up-flow configuration.

Types of Wastewater	Operating OLR range (Kg COD/m3/day)	COD removal efficiency (%)	Reactor configuration	Reference
Protein production from extracted				
Sunflower flour effluent	0.6-9.3	80.0-93.3	UF	[Borja et al., 2001]
POME	10.0-40.0	78.0-94.0	UF	[Borja et al., 1995b]
Ice-cream wastewater	3.2-15.6	94.4	UF	[Borja et al., 1995a]
Distillery effluent	6.11-35.09	80.0-92.0	DF	[Sowmeyan et al., 2008]
Brewery wastewater	0.5-70.0	80.0-90.0	DF	[Alvarado et al., 2008]
Real textile wastewater	0.4-5.0	78.0-89.0	UF	[Sen et al., 2003]

UF-upward flow; DF-downward/inverse flow.

Table 3. Operating OLR range; COD removal efficiency of various wastewater treatments using fluidized bed reactor

The type of support material in the fluidized bed plays an important role to determine the efficiency of the entire treatment system [Garcia et al., 1998], [Sowmeyan et al., 2008] for both inverse flow and up-flow systems. Studies using fluidized bed to treat ice-cream wastewater showed different COD removal efficiencies when different support materials were used. Researcher [Hawkes et al., 1995] found that fluidized bed using granular activated carbon (GAC) gave about 60% COD removal while [Borja et al., 1995a] obtained 94.4% of COD removal using ovoid saponite. Thus suitable support material needs to be selected to obtain high COD removal efficiency in the system.

2.2.3. Up-flow Anaerobic Sludge Blanket (UASB) reactor

Up-flow anaerobic sludge blanket (UASB) technology, normally referred to as UASB reactor, is a form of anaerobic digester that is used in the treatment of wastewater. UASB was developed by [Lettinga et al., 1980] whereby this system has been successful in treating a wide range of industrial effluents including those with inhibitory compounds. The UASB reactor is a methanogenic (methane-producing) digester that evolved from the anaerobic clar-igester. A similar but variant technology to UASB is the expanded granular sludge bed (EGSB) digester. The underlying principle of the UASB operation is to have an aerobic sludge which exhibits good settling properties [Lettinga, 1995]. So far, UASB has been ap-plied for the treatment of potato wastewater [Kalyuzhnyi et al., 1998], [Lettinga et al., 1980], [Parawira et al., 2006]; domestic wastewater [Barbosa et al., 1989], [Behling et al., 1997]; slaughterhouse wastewater [Sayed et al., 1984]; POME [Borja et al., 1994c]. UASB has a rela-tively simple design where sludge from organic matter degradation and biomass settles in the reactor. Organic matter from wastewater that comes in contact with sludge will be di-gested by the biomass granules. Table 4 shows some performances of wastewater treatment using UASB system. For potato wastewater treatment [Kalyuzhnyi et al., 1998] and [Para-wira et al., 2006] both observed foaming and sludge floatation in the UASB reactor when op-erating at higher OLRs (> 6.1kg COD/m3 day). The ability of UASB to tolerate higher OLR for potato wastewater investigated by [Lettinga et al., 1980] compared due to the fact that the latter two studies were conducted at laboratory scale. In general, UASB is successful in COD removal of more than 60% for most wastewater types except for ice-cream wastewater. Researcher [Hawkes et al., 1995] suggested that the lower COD removal percentage from ice-cream wastewater was due to design faults in the reactor's three phase separator and high contents of milk fat that were hard to degrade.

Types of Wastewater	Operating OLR range (Kg COD/m3/day)	COD removal efficiency (%)	Methane Composition (%)	Reference
POME single-stage two-stage	1.8-13.9	63.0-81.0	54.0-67.0	[Kalyuzhnyi et al 1998]
(based on methanogenic reactor)	1.5-6.1	92.0-98.0	59.0-70.0	[Parawira et al. 2006]
Domestic sewage	3.76	74.0	69.0	[Barbosa et al. 1989]
Ice-cream wastewater	0.5-50	50.0	69.6	[Hawkes et al. 1995]
Sugar – beet	4.0-5.0	95.0	N/A	[Lettinga et al. 1980]
Pharmaceutical wastewater	0.27-2.0	26.0-69.0	N/A	[Stronach et al. 1987]
Slaughter wastewater	7.0-11.0	55.0-85.0	65.0-75.0	[Sayed et al. 1984]
Confectionary wastewater	1.25-2.25	66.0	N/A	[Forster et al. 1983]

N/A – data unavailable.

Table 4. Performance of UASB in various wastewater treatments

POME treatment has been successful with UASB reactor, achieving COD removal efficiency up to 98.4% with the highest operating OLR of 10.63 kg COD/m3 day [Borja et al., 1994c]. However, reactor operated under overload conditions with high volatile fatty acid content became unstable after 15 days. Due to high amount of POME discharge daily from milling process, it is necessary to operate treatment system at higher OLR. UASB reactor is advantageous for its ability to treat wastewater with high suspended solid content [Fang et al. 1994]; [Kalyuzhnyi et al., 1998] that may clog reactors with packing material and also provide higher methane production [Kalyuzhnyi et al., 1996]; [Stronach et al., 1987]. However, this reactor might face long start-up periods if seeded sludge is not granulated.

2.2.4. Anaerobic contact digester

The anaerobic contact process is a type of anaerobic digester. Anaerobic digesters are the aerobic equivalents of activated sludge process and are currently used for treating effluents from sugar processing, distilleries, citric acid and yeast production, industries producing canned vegetables, pectin, starch, meat products, etc. This process has been implemented in POME [Ibrahim et al., 1984]; ice-cream wastewater, alcohol distillery wastewater [Vlissidis et al., 1993] and fermented olive mill wastewater treatment [Hamdi et al. 1991]. Concentrated wastewaters are suitable to be treated by anaerobic contact digestion since relatively high quality effluent can be achieved [Jr et al., 1999]. In the study of fermented olive mill wastewater treatment, anaerobic contact was capable of reaching steady state more quickly compared to anaerobic filter; however, more oxygen transfer in the digester (due to mixing) causes this process to be less stable.

2.2.5. Continuous Stirred Tank Reactor (CSTR)

CSTR run at steady state with continuous flow of reactants and products; the feed assumes a uniform composition throughout the reactor, exit stream has the same composition as in the tank. The mechanical agitator provides more area of contact with the biomass thus improving gas production. In POME treatment, CSTR has been applied by a mill under Keck Seng (Malaysia) Berhad in Masai, Johor and it is apparently the only one which has been operating continuously since early 1980's [Tong et al., 2006]. Other applications of CSTR on wastewater treatment include dilute dairy wastewater [Chen et al., 1996]; jam wastewater [Mohan et al., 2008] and coke wastewater [Vazquez et al., 2006].

The CSTR in Kek Seng's Palm oil mill has COD removal efficiency of approximately 83% and CSTR treating dairy wastewater has COD removal efficiency of 60%. In terms of methane composition in generated biogas, it was found to be 62.5% for POME treatment and 22.5-76.9% for dairy wastewater treatment.

2.2.6. Anaerobic contact digestion

Presently there are three categories of anaerobic treatment systems. The first category is the conventional anaerobic digester, which includes two basic designs and another that combines the two. The standard rate digester is the most basic treatment system. It mixes the waste is solely by the movement of gas up through the solid matter and into the top of the tank; there is no external mixing. This process is highly inefficient, for it utilizes only 50 percent of the total waste volume, and requires a very long solid retention time (SRT), usually greater than 30 days; this process has been implemented in POME [Ibrahim et al., 1984]; ice-cream wastewater, alcohol distillery wastewater [Vlissidis et al., 1993] and fermented olive mill wastewater treatment [Hamdi et al., 1991]. Concentrated wastewaters are suitable to be treated by anaerobic contact digestion since relatively high quality effluent can be achieved [Jr et al., 1999]. In the study of fermented olive mill wastewater treatment, anaerobic contact was capable of reaching steady state more quickly compared to anaerobic filter; however, more oxygen transfer in the digester (due to mixing) causes this process to be less suitable.

2.2.7. Membrane separation anaerobic treatment process

Membrane separation has been considered for anaerobic reactors but the technology is still in a development stage. Several studies on membrane anaerobic processes for the treatment of various wastewaters including POME [Fakhru'l et al., 1999] have been performed [Fakhru'l et al., 1994]; [Nagano et al., 1992]; [Pillay et al., 1994]. For example, an ultrafiltration (UF) membrane with a molecular cut-off (MWCO) of 200,000 was used by [76] for biomass/effluent separation in conjunction with an anaerobic process for the treatment of POME. A lower operating pressure (1.5-2 bars) but a higher cross-flow velocity (2.3 m/s) was applied in this study in order to control fouling and to reduce solid deposition on the membrane surfaces. A high COD removal could be obtained in the membrane anaerobic system (MAS), but the permeate displayed a high color content with a low turbidity (less than 10 NTU), including that the color was due to dissolved solids with molecular weights lower than 200,000 g/mol. The particular organics retained in the reactor could be liquefied and decomposed because of the long solid retention time, which was independent of the HRT. The HRT was mainly influenced by the UF membrane flux rates which directly determined the volume of influent that could be fed to the reactor.

2.3. Clean Development Mechanism (CDM)

The Kyoto Protocol is an international agreement linked to the United Nations Framework Convention on Climate Change. The major feature of the Kyoto Protocol is that it sets binding targets for 37 industrialized countries and the European community for reducing greenhouse gas (GHG) emissions.These amount to an average of five per cent against 1990 levels over the five-year period 2008-2012.

The Clean Development Mechanism (CDM), defined in Article 12 of the Protocol, allows a country with an emission-reduction or emission-limitation commitment under the Kyoto Protocol (Annex B Party) to implement an emission-reduction project in developing countries. Such projects can earn saleable certified emission reduction (CER) credits, each equivalent to one tonne of CO2, which can be counted towards meeting Kyoto targets.

The mechanism is seen by many as a trailblazer. It is the first global, environmental investment and credit scheme of its kind, providing a standardized emission offset instrument,

CERs. Besides helping to reduce carbon emission to the environment, CDM has the advantage to offer developing countries such as Malaysia to attract foreign investments to sustain renewable energy projects [Menon, 2002]. Thus, palm oil mills could earn carbon credits as revenue by the utilization of methane gas as renewable energy from anaerobic digestion of palm oil mill effluent. There is a lot of attention has been give to develop anaerobic treatment for POME since the implementation of CDM.

2.4. Comparison of various anaerobic treatment methods in POME treatment

Table 5 shows the performance of several of anaerobic digestion or treatment methods under both mesophilic and thermophilic conditions of POME. As can be seen from Table 5, the fluidized bed reactor has the ability to treat POME at very high organic loading rates; OLR with a short retention time, biogas capture is not emphasized using this process. Therefore, it can be concluded that USFF currently gives the best performance in POME treatment, achieving high COD removal efficiency and high OLR methane production at relatively short hydraulic retention time, HRT compared to conventional and other available anaerobic treatment methods.

retention (days)	Operating OLR (Kg COD/m³/day)	COD removal efficiency (%)	Hydraulic time	Methane composition (%)	Reference
Anaerobic pond 40	1.4		97.8	54.4	[Perez et al., 2001]
Anaerobic digester	2.16	80.7	20	36	[Yacop et al., 2005]
Anaerobic filtration	4.5	94.0	15	63	[Borja et al., 1994b]
Fluidized bed	40.0	78	0.25	N/A	[Borja et al., 1995b]
UASB	10.63	98.4	4	54.2	[Borja et al., 1994c]
UASFF	11.58	97	3	71.9	[Najafpour et al., 2006]
CSTR	3.33	80	18	62.5	[Tong et al., 2006]
Anaerobic contact process[a]	3.44	93.3	4.7	63	[Ibrahim et al., 1984]

N/A: data unavailable.

a In terms of BOD.

Table 5. Performance of various anaerobic treatment methods on POME treatment

Table 6 shows the advantages and disadvantages of each anaerobic treatment method. It can clearly seen that conventional methods are lacking in terms of treatment time, area required for treatment and facilities to capture biogas. However, these methods are more economically viable and have the capacity to tolerate a wider range of OLR. High-rate bioreactors are more effective in biodegradation as shorter retention times are needed, producing higher methane yield while compromising the OLR, capital and operating cost.

Treatment processes	advantages	disadvantages	References
Ponding system	Reliable and stable Anaerobically digested POME from the ponds could be used to culture algae. Cheap, simple to construct, low maintenance costs, the energy needed to operate a ponding system is minimal.	large area of land are required, making it unsuitable for factories located in the near urban and other developed areas. no facilities to capture biogas long retention time.	[Chooi et al. 91984]
Anaerobic filtration	Small reactor volume Producing high quality effluent, short hydraulic times, able to tolerate shock Loadings, retain high biomass Concentration in the packing	lower methane emission, Clogging at high OLRs, High media and support cost Unsuitable for high suspended solid wastewater	[Borja et al., 1994b]
Fluidized bed	Most compact of all high-rate Processes, very well mixed Conditions in the reactor, large Surface area for biomass Attachment	high power requirements for bed fluidization, high cost of carrier media, not suitable for high suspended solid wastewaters	[Jr et al., 1997]
UASB	Useful for treatment of high suspended solid wastewater Producing high quality effluent No media required (less cost)	Performance dependent on sludge settleability, foaming and sludge floatation at high OLRs, long start up period if granulated, seed sludge is not used	[Lettinga, 1995]
UASFF	Higher OLRs achievable compared to operating UASB or anaerobic filtration alone, problems of clogging eliminated Higher biomass retention, more Stable operation, ability to tolerate Shock loadings, suitable for diluted Wastewater.	Granulation inhibition at high volatile fatty acid concentration Lower OLRs when treating suspended solid wastewaters	[Ayati et al., 2006]
CSTR	Provides more contact of wastewater with biomass through mixing, increased gas production compared to conventional method	Less efficient gas production at treatment volume. Less biomass retentio	Hamdi et al., 1991]

Table 6. Advantages and disadvantages of various treatment processes for POME

2.5. Factors influencing anaerobic digester performance

Biogas coming from biomethanization or anaerobic digestion represents an attractive strategy for both biomass waste treatment and recycling and is of great interest from an environmental point of view and may benefit society by providing a clean fuel source from renewable energy. This technology is accomplished by a series of biochemical transformations, which can be toughly separated into a first step where hydrolysis, acidification and liquefaction take a place and second step where acetate, hydrogen and carbon dioxide are transformed into biogas with methane content between 60-80%, which cover a large part of energy. Many factors govern the performance of anaerobic digesters where adequate control is required to prevent reactor failure. These factors are operating temperature, pH, mixing, nutrients for bacteria and organic loading rates into the digester.

2.5.1. Operating temperature

One of the most important factors affecting anaerobic digestion of organic waste is temperature. Anaerobic digestions can be developed at different temperature ranges including mesophilic temperatures (approximately 35ºC) and thermophilic temperatures ranging from 55 ºC to 60 ºC. Conventional anaerobic digestion is carried out at mesophilic temperatures (35–37 ºC), mainly because of the lower energy requirements and better stability of the process. POME is discharged at temperatures around 80-90 oC [Zinatizadeh et al., 2006] which actually makes treatment at both mesophilic and thermophilic temperature feasible especially in tropical countries like Malaysia. Effect of temperature on the performance of anaerobic digestion was investigated by [H-Q et al., 2002a] and found that substrate degradation rate and biogas production rate at 55 oC was higher than operation at 37 oC. Studies have reported that thermophilic digesters are able to tolerate higher OLRs and operate at shorter HRT while producing more biogas [Ahn et al., 2002], [Kim et al., 2006], and [Yilmaz et al., 2008]. However, failure to control temperature increase can result in biomass washout [Lau et al., 1997] with accumulation of volatile fatty acid due to inhibition of methanogenesis. At high temperatures, production of volatile fatty acid is higher compared to mesophilic temperature range [H-Q et al., 2002a].

2.5.2. pH

A pH (potential of Hydrogen) measurement reveals if a solution is acidic or alkaline (also base or basic). If the solution has an equal amount of acidic and alkaline molecules, the pH is considered neutral. The microbial communities in anaerobic digesters are sensitive to pH changes and methanogens are affected to a great extend [Jr et al., 1999]. Several cases of reactor failure reported in studies of wastewater treatment are due to accumulation of high volatile fatty acid concentration, causing a drop in pH which inhibited methanogenesis Parawira et al. (2006), [Patel et al., 2002]. Thus, volatile fatty acid concentration is an important parameter to monitor to guarantee reactor performance [Buyukkamaci et al., 2004]. It was found that digester could tolerate acetic acid concentrations up to 4000 mg/l without inhibition of gas production Stafford (1982). To control the level of volatile fatty acid in the system, alkalinity has to be maintained by recirculation of treated effluent [Najafpour et al., 2006], [Borja et al., 1996a] to the digester or addition of lime and bicarbonate salt [Gerardi, 2003].

2.5.3. Mixing

Distribution of bacteria, substrate, nutrients and temperature equalization by means of adequate mixing, are known to be crucial for the overall anaerobic digester (AD) process [Chapman, 1989]. Several investigations show that improvements in reactor performance can be achieved when changes in mixing intensity are imposed [Angelidaki et al., 2004]. According to [Gerardi, 2003] the main advantages of mixing in AD are: minimization of solids accumulation that may restrict reactor hydraulics, reduction of scum build up, elimination of temperature stratification and maintaining close contact between substrate particles and microbial communities. In a sequential experiment [Stroot et al., 2001] studied the feasibility of co-digestion of municipal solid waste, primary sludge and waste activated sludge (WAS) under mesophilic conditions in laboratory scale continuous stirred tank reactors (CSTRs) under different OLR and mixing conditions.

2.5.4. Organic loading rates

Organic loading rate is defined as the application of soluble and particulate organic matter. It's typically expressed on an area basis as pounds of BOD per unit area. Various studies have shown that higher OLRs will reduce COD removal efficiency in wastewater treatment systems [Torkian et al., 2003], [Sanchez et al., 2005]. However, gas production will increase with OLR until a stage when methanogens could not work quick enough to convert acetic acid to methane.

3. Conclusions

The performance of anaerobic treatment for POME and effects of organic loading rates were thoroughly reviewed. The palm oil industry is an indisputable source of pollution in Malaysia. In order to counteract the negative impact of this source, anaerobic digestion is an advantageous method for POME treatment as it generates valuable and product that can be exchanged into revenue when registered as a clean development mechanism CDM project. Furthermore, research can be done to develop a thermophilic anaerobic bioreactor with minimal control to ease system operation. Moreover, intensity of mixing in the thermophilic range should be investigated to obtain an optimum mixing rate that will keep microbial consortia in close proximity and at the same time improve the system efficiency. Furthermore, operation costs can be reduced through utilization of biogas for heat or electricity energy generation in the plant.

Author details

N.H. Abdurahman[1,2], Y.M. Rosli[1,2] and N.H. Azhari[1,2]

1 LebuhrayaTun Razak, Gambang, Kuantan, Malaysia

2 Faculty of Chemical and Natural Resources Engineering, University Malaysia Pahang, Malaysia

References

[1] Ahmad, A. L, Chong, M. F, Bhatia, S, & Ismail, S. (2006). Drinking water reclamation from palm oil Mill effluent (POME) using membrane technology. *Desalination*, 191, 35-44.

[2] Ahmad, A. L, Chong, M. F, & Bhatia, S. (2007). Mathematical modeling of multiple solutes system for reverse osmosis process in palm oil mill effluent (POME) treatment. *Chemical Engineering Journal*, 132, 183-193.

[3] Ayati, B, & Ganjidoust, H. (2006). Comparing the efficiency of UAFF and UASB with hybrid reactor in treating wood fiber wastewater. Iranian Journal of Environmental HealthScience Engineering :, 3, 39-44.

[4] Alvarado-lassman, A, Rustrian, E, Garcia-alvarado, M. A, Rodriguez, G. C, & Houbron, E. (2008). Brewery wastewater treatment using anaerobic inverse fluidized bed reactors. Bioresource Technology: , 99, 3009-3015.

[5] Ahn, J, & Foster, H. . , C. F. (2002). A comparison of mesophilic and thermophilic anaerobic up-flow filters treating paper-pulp-liquors. Process Biochemistry: , 38, 257-262.

[6] Angelidaki, I, & Sanders, W. (2004). Assessment of the anaerobic biodegradability of macropollutants. *Reviews in Environmental Science and Biotechnology*, 3, 117-129.

[7] Ma, A. N, Cheah, S. C, & Chow, M. C. (1993). Current status of palm oil processing wastes management. In: Waste Management in Malaysia: Current status and Prospects for Bioremediation, , 111-136.

[8] Ma, A. N. (1999). Treatment of palm oil mill effluent. Oil Palm and Environment: Malaysian Perspective. Malaysia Oil Palm Growers' Council. (1999), , 277.

[9] Borja, R, Gonzalez, E, Raposo, F, Millan, F, & Martin, A. (2001). Performance evaluation of a Mesophilic anaerobic fluidized-bed reactor treating wastewater derived from the production of proteins from extracted sunflower flour. Bioresource Technology: , 76, 45-52.

[10] Borja, R, Banks, C. J, & Sanchez, E. (2006a). Anaerobic treatment of palm oil mill effluent in a two- stage up-flow anaerobic sludge blanket (UASB) reactor. Journal of Biotechnology , 45, 125-135.

[11] Borja, R, & Banks, C. J. (1994c). Anaerobic digestion of palm oil mill effluent using an up-flow Anaerobic sludge blanket reactor. Bimass and Bioenergy: , 6, 381-389.

[12] Borja, R, Banks, C. J, & Sanchez, E. (1996a). Anaerobic treatment of palm oil mill effluent in a two- stage up-flow anaerobic sludge blanket (UASB) reactor. Journal of Biotechnology , 45, 125-135.

[13] Borja, R, & Banks, C. J. (1994a). Anaerobic digestion of palm oil mill effluent using an up-flow Anaerobic Sludge blanket reactor. Biomass and Bioenergy , 6, 381-389.

[14] Borja, R, & Banks, C. J. (1994b). Treatment of palm oil mill effluent by up-flow anaerobic filtration. Journal of Chemical Technology and Biotechnology , 61, 103-109.

[15] Borja, R, & Banks, C. J. (1995a). Response of anaerobic fluidized bed reactor treating ice-cream wastewater to organic, hydraulic, temperature and pH shocks. *Journal of Biotechnology*, 39, 251-259.

[16] Borja, R, & Banks, C. J. (1995b). Comparison of anaerobic filter and an anaerobic fluidized bed reactor treating palm oil mill effluent. *Process Biochemistry*, 30, 511-521.

[17] Bodkhe, S. (2008). Development of an improved anaerobic filter for municipal wastewater treatment. Bioresource Technology: , 99, 222-226.

[18] Behling, E, Diaz, A, Colina, G, Herrera, M, Gutierrez, E, Chacin, E, Fernandez, N, & Forster, C. F. (1997). Domestic wastewater treatment using UASB reactor. Bioresource Technology: , 61, 239-245.

[19] Buyukkamaci, N, & Filibeli, A. (2004). Volatile fatty acid formation in anaerobic hybrid reactor. *Process Biochemistry*, 39, 1491-1494.

[20] Chen, T, & Shyu, H. , W. -H. (1996). Performance of four types of anaerobic reactors in treating very dilute dairy wastewater. *Biomass and Bioenergy*, 11, 431-440.

[21] Chong, M. F. simulation and design of membrane based palm oil mill effluent (POME) treatment plant from pilot plant studies (2007PhD. Thesis. Univer. siti Sains Malaysia.

[22] Chooi, C. F, Ponding system for palm oil mill effluent treatment. PORIM 9, ((1984).

[23] Chin, K. K. (1981). Anaerobic treatment kinetics of palm oil sludge. *Water Research*, 199 EOF-202 EOF.

[24] Cote, C, Daniel, I. M, & Quessy, S. (2006). Reduction of indicator and pathogenic microorganisms by psychrophilic anaerobic digestion in swine slurries. *Bioresource Technology*, 97, 686-691.

[25] Chapman, D. (1989). Mixing in anaerobic digesters: state of art. In: P. Cheremisinoff (Ed.) Encyclopedia of environmental control technology, Houston, TX: Gulf, , 325-354.

[26] Doble, M, & Kumar, A. (2005). Biotreatment of Industrial Effluents. Elsevier Butterworth- Heinemann, Oxford, United Kingdom, , 19-38.

[27] Fakhru, l-R. a. z. i. A. ((1994). Ultrafiltration membrane separation for anaerobic wastewater treatment. Water.Sci. Technol , 30(12), 321-327.

[28] Fakhru, l-R. a. z. i, & Noor, A. , M. J. M. M. (1999). Treatment of palm oil mill effluent (POME) with the Membrane anaerobic system (MAS). Water Sci. Technol: , 39, 159-163.

[29] Farhadian, M, Borghei, M, & Umrania, V. V. Treatment of beet sugar water by UAFB bioprocess. Bioresource Technology ((2007).

[30] Forster, C. F, & Wase, D. A. J. (1983). Anaerobic treatment of dilute wastewaters using an upflow sludge blanket reactor. Environmental Pollution (Series A): , 31, 57-66.

[31] Fang, H. H. P, & Chui, H. K. (1994). Comparison of startup performance of four anaerobic reactors for The treatment of high-strength wastewater. ResourcesConservation and Recycling (1994): 11, , 123 EOF.

[32] Gangagni RaoA., G. Venkata Naidu., K. Krishna Prasad., N. Chandrasekhar Rao., N. Venkata Mohan., A. Jetty., P.N. Sarma. ((2005). Anaerobic treatment of wastewater with high suspended solids from a bulk drug industry using fixed film reactor (AFFR). Bioresource Technology: , 96, 87-93.

[33] Gee, P. T, & Chua, N. S. (1994). Current status of palm oil mill effluent by watercourse discharge. In: PORIM National Palm Oil Milling and Refining Technology Conference.

[34] Guerrero, L, Omil, F, Mendez, R, & Lema, J. M. (1999). Anaerobic hydrolysis and acidogenesis of wastewater from food industries with high content of organic solids and protein. Water Resource Journal , 33, 3281-3290.

[35] Gerardi, M. H. The microbiology of Anaerobic Digesters. Wiley-Interscience, New Jersey ((2003). , 51-57.

[36] Gopal, G, & Ma, A. N. (1986). The comparative economics of palm oil mill effluent treatment and resource recovery systems. National Workshop on Recent Developments in Palm Oil Milling Technology & Pollution Control.

[37] Garcia-calderon, D, Buffiere, p, Moeltta, R, & Elmaleh, S. (1998). Anaerobic digestion of wine Distillery wastewater in down-flow fluidized bed. *Water Research*, 32, 3593-3600.

[38] Hamdi, M, & Garcia, J. L. (1991). Comparison between anaerobic filter and anaerobic contact process for fermented olive mill wastewaters. Bioresource Technology: , 38, 23-29.

[39] Hawkes, F. R, Donnelly, T, & Anderson, G. K. (1995). Comparative performance of anaerobic Digesters operating on ice-cream wastewater. *Water Research*, 29, 525-533.

[40] Yu, H-Q, Hu, Z-H, Hong, T. Q, & Gu, G. W. (2002a). Performance of an anaerobic filter treating soybean processing wastewater with and without effluent recycle. *Process Biochemistry*, 38, 507-513.

[41] Ibrahim, A, Yeoh, B. G, Cheah, S. C, Ma, A. N, Ahmad, S, Chew, T. Y, Raj, R, & Wahid, M. J. A. (1984). Thermophilic anaerobic contact digestion of palm oil mill effluent. Water Science and Technology , 17, 155-165.

[42] Jr, G, Leslie, G. T, & Daigger, H. C Lim. ((1999). Biological Wastewater Treatment. second ed. CRC Press. Revised & Expanded.

[43] Jawed, M, & Tare, V. (2000). Post-mortem examination and analysis of anaerobic filters. Bioresource Technology: , 72, 75-84.

[44] Kim, J. K, Oh, B. R, Chun, Y. N, & Kim, S. W. (2006). Effects of temperature and hy-draulic retention time on anaerobic digestion of food waste. *Journal of Bioscience and Bioengineering*, 102, 328-332.

[45] Kalyuzhnyi, S, & De Los, E. L. Santos., J.R, Martinez. Anaerobic treatment of raw and preclarified potato-maize wastewater in a UASB reactor. Bioresource Technology ((1998).

[46] Kalyuzhnyi, S. V, Skylar, V. I, Davlyatshina, M. A, Parshina, S. N, Simankova, M. V, Kostrikina, N. A, & Nozhevnikova, A. N. (1996). Organic removal and microbiologi-cal features of UASB- reactor under various organic loading rate. s. Bioresource Tech-nology: , 55, 47-54.

[47] Lettinga, G, Velson, A. F. M, Homba, S. W, De Zeeuw, W, & Klapwiki, A. (1980). Use of up-flow Sludge blanket (USB) reactor concept for biological wastewater treatment, especially for anaerobic treatment. Biotechnology and Bioengineering: , 22, 699-734.

[48] Lettinga, G. (1995). Anaerobic digestion and wastewater treatment systems. Antonie Van Leeuwenhoek: , 67, 3-28.

[49] Lau, I. W. C, & Fang, H. H. P. (1997). Effect of temperature shock to thermophilic granules. *Water Research*, 31, 2626-2632.

[50] Leal, K, Chachin, E, Gutierez, E, Fernandez, N, & Forster, C. F. (1998). A mesophilic digestion of brewery wastewater in an unheated anaerobic filter. Bioresources Tech-nology: , 65, 51-55.

[51] Liu, W. T, Chan, O. C, & Fang, H. H. P. (2002). Microbial community dynamics dur-ing start-up of Acidogenic anaerobic reactors. Water Resource: , 36, 3203-3210.

[52] Linke, B. (2006). Kinetic study of thermophilic anaerobic digestion of solid wastes from potato processing. Biomass and Bioenergy , 30, 892-896.

[53] MetcalfEddy. Wastewater Engineering Treatment and Reuse,. fourth ed. McGraw Hill ((2003). , 96-97.

[54] Mchugh, S, Reilly, C. O, Mahony, T, Colleran, E, & Flaherty, V. O. (2003). Anaerobic granular Sludge bioreactor technology. *Reviews in Environmental Science and Biotech-nology*, 2, 225-245.

[55] Menon, R. N. Carbon credits and clean development mechanism. Palm Oil Engineer-ing Bulletin ((2002).

[56] Mohan, S, & Sunny, N. (2008). Study of biomethanization of waste water from jam industries. Bioresource Technology: , 99, 210-213.

[57] Ma, A. N, Tajima, Y, Asahi, M, & Hanif, J. (1997). Effluent treatment-evaporation method. PORIM Engineering News , 44, 7-8.

[58] Nagano, A, Arikawa, E, & Kobayashi, H. (1992). The treatment of liquor wastewater containing gihg-strength suspended solids by membrane bioreactor system. Water Sci. Technol: , 26, 887-895.

[59] Najafpour, G. D, Zinatizadeh, A. A. L, Mohamed, A. R, & Hasnain, M. Isa., H.Nasrol-lahzadeh. ((2006). High-Rate anaerobic digestion of palm oil mill effluent in an up-flow anaerobic sludge-fixed film bioreactor. Biochemistry (2006): 41, , 370 EOF.

[60] Nebot, E, Romero, L. I, Quiroga, J. M, & Sales, D. (1995). Effect of the feed frequency on the Performance of anaerobic filters. *Anaerobe*, 1, 113-120.

[61] Patel, H, & Madamwar, D. (2002). Effects of temperature and organic loading rates on biomethanation of acidic petrochemical wastewater using an anaerobic upflow fixed-film reactor. Bioresource Technology (2002): 82, , 65 EOF-71 EOF.

[62] Pillay, V. L, Townsend, B, & Buckley, C. A. (1994). Improving the performance of anaerobic digesters at wastewater treatment works: the coupled cross-flow microfil-tration/digester process. Water Sci. Technol: , 30, 329-337.

[63] Perez, M, Romero, L. I, & Sales, D. Organic matter degradation kinetics in an anaero-bic thermophilic fluidized bed bioreactor, Anaerobic ((2001).

[64] Perez, M, Romero, L. I, & Sales, D. (1998). Comparative performance of high rate anaerobic thermophilic technologies treating industrial wastewater. *Water Research*, 32, 559-564.

[65] Perez, M, Rodriguez-cano, R, Romero, L. I, & Sales, D. (2007). Performance of anaero-bic thermophilic fluidized bed in the treatment of cutting-oil wastewater. Bioresource Technology , 98, 3456-3463.

[66] Reyes, O, Sanchez, E, Rovirosa, N, Borja, R, Cruz, M, Colmenarejo, M.F, Escobedo, R, Ruiz, M, Rodriguez, X, & Correa, O. . ((1999).). Low-strength wastewater treatment by a multistage

[67] Russo, C, & Jr, G. L. San't Anna., S.E. De Carvalho. ((1985). An Aerobic filter applied to the treatment of distillery wastewaters. Agricultural Wastes: Anaerobic filter packed with waste tyre rubber. Bioresource Technology 70, 55-60., 14, 301-313.

[68] Ruiz, I, Veiga, M. C, De Santiago, P, & Blazquez, R. (1997). Treatment of slaughter-house wastewater in a USAB reactor and an anaerobic filter. Bioresource Technolo-gy: , 60, 251-258.

[69] Rbarbosa, R. A, & Sant, G. L. Anna Jr. ((1989). Treatment of raw domestic sewage in an UASB reactor. Water Research: , 23, 1483-1490.

[70] Stronach, S. M, Rudd, T, & Lester, J. N. (1987). Start-up of anaerobic bioreactors on high strength Industrial wastes. *Biomass*, 13, 173-197.

[71] Stroot, P. G, Mcmahon, K. D, Mackie, R. I, & Raskin, L. (2001). Anaerobic co-diges-tion of municipal solid waste and biosolids under various mixing conditions-I. Di-gester performance. Water Research: , 35, 1804-1816.

[72] Saravanane, R, Murthy, D. V. S, & Krishnaiah, K. Treatment of anti-osmotic drug based pharmaceutical effluent in an upflow anaerobic fluidized bed system. Waste Management ((2001). , 563 EOF-8 EOF.

[73] Sen, S, & Demirer, G. N. (2003). Anaerobic treatment of real textile wastewater with a fluidized bed reactor. *Water Research*, 37, 1868-1878.

[74] Sanchez, E, Borja, R, Travieso, L, Martin, A, & Colmenarejo, M. F. (2005). Effect of organic loading rate on the stability, operational parameters and performance of a secondary upflow anaerobic sludge bed reactor treating piggery waste. *Bioresource Technology*, 96, 335-344.

[75] Stafford, D. A. (1982). the effects of mixing and volatile fatty acid concentrations on anaerobic Digester performance. Biomass, 2, 43-55.

[76] Sayed, S, De Zeeuw, W, & Lettinga, G. (1984). Anaerobic treatment of slaughterhouse waste using a flocculent sludge UASB reactor. Agricultural Wastes: , 11, 197-226.

[77] Sowmeyan, R, & Swaminathan, G. (2008). Evaluation of inverse anaerobic fluidized bed reactor for Treating high strength organic wastewater. Bioresource Technology: , 99, 3877-3880.

[78] Toldra, F, Flors, A, Lequerica, J. L, & Valles, S. (1987). Fluidized bed anaerobic biodegradation of food industry wastewaters. *Biological Wastes*, 21, 55-61.

[79] Torkian, A, Eqbali, A, & Hashemian, S. J. (2003). The effect of organic loading rate on the performance of UASB reactor treating slaughterhouse effluent. Resources Conservation & Recycling: , 40, 1-11.

[80] Tong, S. L, & Jaafar, A. B. (2006). POME Biogas capture, upgrading and utilization. Palm Oil Engineering Bulletin: , 78, 11-17.

[81] Tay, J. H. Complete reclamation of oil palm wastes. ResourcesConservation and Recycling 5, ((1991). , 383 EOF-392 EOF.

[82] Ulrich, G. D, & Vasudevan, P. T. (2004). Chemical Engieering Process Design and Economics: A Practical Guide, second ed. Process Publishing Company.

[83] Vlissidis, A, & Zouboulis, A. I. (1993). Thermophilic anaerobic digestion of alcohol distillery wastewaters. *Bioresource Technology*, 43, 131-140.

[84] Vazquez, I, Rodriguez, J, Maranon, E, Castrillon, L, & Fernandez, Y. (2006). Simultaneous removal of phenol, ammonium and thiocyanate from coke wastewater by aerobic degradation. Journal of Hazardous MaterialsB: , 137, 1773-1780.

[85] Van Der Merwe, M, & Britz, T. J. (1993). Anaerobic digestion of baker's yeast factory efficient using an anaerobic filter and hybrid digester. Bioresource Technology: , 43, 169-174.

[86] Parawira, W, Murto, M, Zvauya, R, & Mattiasson, B. (2006). Comparative performance of a UASB reactor and an anaerobic packed-bed reactor when treating potato
waste leachate. *Renewable Energy*, 31, 893-903.

[87] Wang, Z, & Banks, C. J. (2007). Treatment of a high-strength substrate-rich alkaline
leachate using an Aerobic filter. Waste Management , 27, 359-366.

[88] Yilmaz, T, Yuceer, A, & Basibuyuk, M. A comparison of the performance of mesophilic and thermophilic anaerobic filters treating papermill wastewater. Bioresource
Technology ((2008). , 156 EOF-163 EOF.

[89] Yacob, S, Hassan, M. A, Shirai, Y, Wakisaka, M, & Subash, S. (2006a). Baseline study
of methane Emission from anaerobic ponds of palm oil mill effluent treatment. *Science of the Total Environment*, 366, 187-196.

[90] Yacop, S, Hassan, M. A, Shirai, Y, Wakisaka, M, & Subash, S. (2005). Baseline study
of methane Emission from open digesting tanks of palm oil mill effluent treatment.
Chemosphere , 59, 1575-1581.

[91] Yacob, S, Shirai, Y, Hassan, M. A, Wakisaka, M, & Subash, S. operation of semi-com-
mercial closed anaerobic digester for palm oil mill effluent treatment. Process Biochemistry, 41, 962-964.

[92] Zinatizadeh, A. A. L, Mohamed, A. R, Abdullah, A. Z, Mashitah, M. D, & Hasnain,
M. Isa., G.D, Najafpour. ((2006). Process modeling and analysis of palm oil mill effluent treatment in an up-flow anaerobic sludge fixed film bioreactor using response
surface methodology (RSM). Water Research, 40, 3193-3208.

Chapter 4

Impact of Industrial Water Pollution on Rice Production in Vietnam

Huynh Viet Khai and Mitsuyasu Yabe

Additional information is available at the end of the chapter

1. Introduction

Vietnam has achieved the average GDP growth rate of 6.71% per year. The industrial sector has mainly contributed economic development in Vietnam, with annual growth of 12% during the period of 200-2009. In line with its industrialization and modernization policies, Vietnam has rapidly changed economic structure from agriculture base to industrial economy. The industrial and construction sector only contributed 26 percent of national GDP in 1986, but it rapidly increases to 40.3 percent in 2009.

Economic development has brought many benefits to Vietnam. Income, public transportation and, in general, quality of life have gradually improved while the percentage of people below the poverty threshold has reduced. However, there have also been many negative consequences of rapid industrialization, particularly on agriculture and ecosystem health, because of the exploitation of natural resources and pollution. The two biggest cities in Vietnam, Ha Noi and Ho Chi Minh, have been ranked as the worst cities in Asia for dust pollution (The World Bank, 2008). Within Vietnam, Ho Chi Minh, the largest city, is at the top of the national pollution list (The World Bank, 2007). This pollution, into the air, water and land, is released by various, large industries. For instance, footwear manufacturing releases 11% of the air pollution load, 10% of the land pollution load and 6% of the water pollution load, while the plastic products manufacturing industry produces 10, 13 and 9% of the air, land and water pollution load, respectively. The main pollution sources do not necessarily come from the largest industries. The cement industry, which only has 12 factories and employs 0.5% of the provincial workforce, releases 24% of the air pollution load (ICEM, 2007). Similarly, the 160 paper factories employ only 0.8% of the provincial workers but contribute 14% of the water pollution load.

According to the Department of Science, Technology, and Environment of Tay Ninh, since almost all industrial zones have not installed wastewater treatment systems in Vietnam, the existence of industrial wastewater contamination appears almost everywhere. Wastewater from thousands of industrial facilities in 30 industrial areas and from small factories and businesses in the basin is the main source of pollution in the Dong Nai river of Ho Chi Minh City [1]. The untreated wastewater contaminating oil from Hai Au concrete factory has been released directly into paddy fields approximately 200m³/day and 1,500 m³/day for Phuoc Long textile firm. However, it is difficult to know the actual damage and loss due to the contamination of untreated wastewater from industrial activities in Vietnam (Quang, 2001).

There have been a number of empirical agricultural studies concerning environmental problems, such as soil degradation, wind and water erosion in the world; however, few have specifically examined the impact of industrial pollution. Bai (1988) conducted field experiments in wheat lands irrigated with wastewater from the Liangshui River, the Tonghui River and the Wanquan River. He reported that wastewater irrigation caused a reduction in wheat yield by 8–17.1%. Similar studies in the Geobeidian area of the Tonghui River and the Yizhuang area of the Lianghe River reported that yields of wheat and rice cultivated in unpolluted soils in the sewage-irrigated area decrease by about 10% of the yields obtained in clean water-irrigated areas. In the sewage-irrigated area with polluted soils, yields of wheat and rice grown reduce by 40.6% and 39% of those in clean irrigation areas.

Chang *et al.* (2001) analyzed the impact of industrial pollution on agriculture, human health and industrial activities in Chongqing. To determine the effect of sewage-irrigation, they proposed expressing yield reductions as a function of the comprehensive water pollution index. Using this approach, reductions in yield due to sewage irrigation were about 10% for wheat and 30% for rice and vegetables. To evaluate the effects of polluted water irrigation, Lindhjem (2007) compared crop quality and quantity between a wastewater-irrigated area and a clean water-irrigated area. The total loss of corn and wheat production was estimated to be RMB 360 per mu, of which RMB 285 was caused by reduction in quantity, and RMB 75 was the reduction in quality. This paper also cites the study of Song (2004) that used dose-response functions to estimate the reductions in quantity and quality of crops from polluted water irrigation. Water pollution decreased rice production by 20% and quality by about 4%.

A study by The World Bank (2007) also used dose-response functions to calculate the economic losses from crop damage caused by water pollution, in terms of both reductions in crop quantity and quality (excess pollutant levels and substandard nutritional value). The economic cost of wastewater irrigation in China was estimated to be about 7 billion RMB annually for the four major crops (wheat, corn, rice, and vegetables). Reddy and Behera (2006) evaluated the impact of water pollution on rural communities in India, in terms of agricultural production, human heath, and livestock, using the effects on production, replacement costs and human capital approaches. The study estimated that the total loss per household per annum due to water pollution was $282.5, of which $213.2 was from agriculture, $16.3 from livestock and $53 from human health.

There has been some studies in recent years on industrial pollution in Vietnam such as the report written by Thong and Ngoc (2004) presented a descriptive analysis of data col-

---

1 The speech of Dr. Trinh Le, the Institute of Tropical Technology and Environmental Protection.

lected from 32 industrial estates in southern Vietnam to determine the factors affecting investment on wastewater treatment plants. It performed that water pollution was a serious problem in the big industrial estates of Ho Chi Minh City, Binh Duong, Dong Nai and Ba Ria-Vung Tau Provinces, and that financial constraints and lack of space were the main reasons why many small and medium-sized enterprises did not invest in wastewater treatment systems. Hung *et al.*(2008) studied the effects of trade liberalization on the environment, using data from the Viet Enterprise Survey of 2002 and the World Bank's Industrial Pollution Projection System. They found that trade liberalization led to greater pollution and environmental degradation but that the Vietnamese people have gradually recognized the importance of environmental protection.

However, because of a lack of information on the costs of pollution, national and local authorities in Vietnam have not paid much attention to pollution control measures. In this study, we review the literature on this topic and estimate the damage of rice production due to water pollution. Our findings could help governmental bodies enforce existing water pollution regulations, for example, TCVN 5945 on water pollution standards or Decree 67 on wastewater pollution charges, also help recognize and understand the failure of some of the current environmental policies in Vietnam. Our study could also provide useful information to authorities, such as the Natural Resources and Environment, and industries to manage water pollution and data for cost-benefit analyses of treatment projects in the industrial zones of Vietnam.

2. Evaluation concept

The total economic loss of rice production includes three factors. First, a reduction in crop quantity assumes that water pollution decreases rice yield. Second, a reduction in rice quality, which is measured as price, assumes that the lower price of rice in a particular region could reflect reduced rice quality due to water pollution. Third, an increase in input costs assumes that farms may attempt to compensate for the possible productivity losses by implementing activities that are capable of offsetting this possible loss but are more costly to implement. The expectation of the profit loss is summarized by the following formula:

$$
\begin{aligned}
\pi_p &= \left(\bar{P} - \Delta P\right)\left(\bar{Q} - \Delta Q\right) - \left(\bar{C} + \Delta C\right) \\
&= \bar{P}\bar{Q} - \bar{P} \times \Delta Q - \Delta P \times \bar{Q} + \Delta P \times \Delta Q - \bar{C} - \Delta C \\
&= \left(\bar{P}\bar{Q} - \bar{C}\right) - \left(\bar{P} \times \Delta Q + \Delta P \times \bar{Q} + \Delta \bar{C}\right) + \Delta P \times \Delta Q \\
&= \quad \pi_n \quad - \quad\quad\quad \text{Profit loss}
\end{aligned}
\tag{1}
$$

$$
\Rightarrow \text{Profit loss} = \bar{P} \times \Delta Q \quad + \quad \Delta P \times \bar{Q} \quad + \quad \Delta \bar{C}
$$
$$
= \text{Quantity loss} + \text{Quality loss} + \text{Cost increase}
\tag{2}
$$

where π_n and π_p are the rice profits in the non-polluted and polluted areas. Because $\Delta P \times \Delta Q$ is small compared with the other parts of the equation, it can be ignored and assumed to be 0.

However, it is complicated to estimate quality loss through the proxy of price because there are many other unobservable factors, excepting water pollution, which affect the price of rice. Thus, the study only calculates the three elements affected by water pollution:

- *Quantity loss:* Water pollution causes a decrease in rice yield. The production function approach is used to estimate the loss of rice yield.

- *Cost increase:* Since farms may aim and indeed be able to compensate for the possible productivity losses by implementing activities which are capable of offsetting this possible loss but are more costly to implement. In such circumstances, because it is not productivity which will be impacted, but production costs, cost function approach is applied to assess the impacts of pollution in economic terms.

- *Profit loss:* This is defined as total loss of net economic return estimated by the comparison of profit functions between two selected areas (one is considered as the polluted, other is the non-polluted area). The difference in rice profits of two regions is considered as total loss of net economic return due to industrial pollution.

3. Empirical model

We surveyed rice farmers in two areas with the assumption that they had the same natural environment conditions and social characteristics, and only differed with respect to pollution. One area was considered to be the polluted area, receiving wastewater from nearby industrial parks, while the other area was assumed to be the non-polluted area, being distant from sources of industrial pollutants. The productivity loss of rice production caused by water pollution was estimated by the difference in rice yield between the two regions (Translog production function approach). The similar calculation was applied for cost increase and profit loss due to water pollution by applying the methods of Cobb-Douglas cost function and translog profit function respectively.

3.1. Production function approach

The production function approach is that industrial activities possibly have a negative impact on the outputs, cost and profit of producers through the effect of environment. Environment affects goods or services existing in the market through the value change of their outputs, for instance, the reduced value of fish caught because of river pollution. The production function approach is often used to estimate the effect of environment change on soil erosion, deforestation, fisheries, the impact of air and water pollution on agriculture and so on (Bateman *et al.*, 2003)

A literature search on the production function approach in rice production Vietnam was conducted to make sure that relevant variables will be included in the farm survey questionnaire and to examine the suitability of existing rice production models for the research. There are a number of studies related to rice production in Vietnam. Kompas (2004) and Linh (2007) used a stochastic production frontier to estimate the technical efficiency of rice production in Vietnam. Do and Bennett (2007) used a production function approach with flood duration and relative location of upstream and downstream farmers variables to estimate the cost of changing wetland management, representing the reduced income of rice production in the Mekong River Delta. The loss of rice productivity was estimated based on the differences in rice yield between upper and lower of the Tram Chim park dyke. The results showed that the rice productivity in the lowering of park dyke decreased 0.06 tons per hectare per annum, which led to the profit loss of VND 0.07 million per hectare per annum. These three studies used the Cobb-Douglas functional form of the rice production function approach. This study uses a translog functional form and does test for checking the existence of Cobb-Douglass. The model takes the basic form:

$$Y = f(L,K,I,Z,E,F) \tag{3}$$

where Y is the rice yield of a farmer in the studied year (tones/ha), L is the number of labors for rice cultivation (man-days/ha), K is capital input (VND/ha), I is a vector of material inputs as seeds (kg/ha), fertilizers (kg/ha), herbicide (ml/ha) and pesticides (ml/ha), Z is a vector of social-economic characteristics of farmers, and E is a vector of farming conditions, and F is the relative location of farms (polluted site = 1, non-polluted site = 0)

The test for the existence of quantity loss due to water pollution is:

$$\begin{aligned} H_0 &: \text{Quantity loss} = 0 \text{ or Coef. of F} = 0 \\ H_1 &: \text{Quantity loss} > 0 \text{ or Coef. of F} < 0 \end{aligned} \tag{4}$$

The reduced yield of rice is defined as the difference in the average rice yield between the non-polluted and polluted site. It is estimated by following equation:

$$\Delta Y = f(\bar{L},\bar{K},\bar{I},\bar{Z},\bar{E},F=0) - f(\bar{L},\bar{K},\bar{I},\bar{Z},\bar{E},\Gamma=1) \tag{5}$$

where ΔY is the average yield loss caused by water pollution (kg/ha); $\bar{L}$, $\bar{K}$, $\bar{I}$, $\bar{Z}$, $\bar{E}$ are the average of labor, capital input, material inputs, social-economic characteristics, and farming conditions, respectively.

As mentioned earlier, a translog functional form is used in the study. The production functional form in the polluted and non-polluted areas is written as followed (Tim & Battese, 2005):

$$\ln(Y) = \alpha_0 + \alpha_1 \ln(L) + \alpha_2 \ln(K) + \alpha_3 \ln(I) + \frac{1}{2}\alpha_{11}\left(\ln(L)\right)^2 + \alpha_{12}\ln(L)\ln(K) + \alpha_{13}\ln(L)\ln(I) +$$
$$+ \frac{1}{2}\alpha_{22}\left(\ln(K)\right)^2 + \alpha_{23}\ln(K)\ln(I) + \frac{1}{2}\alpha_{33}\left(\ln(I)\right)^2 + \sum_{k=1}^{5}\beta_k Z_k + \sum_{h=1}^{4}\delta_h E_h + \gamma F \tag{6}$$

where Y, L, K, I, F are the same as in the above equations and Z_1, Z_2, Z_3, Z_4 are the variables of the gender (1 = male, 0 = female), the age (years), the number of school year (years), attending trainings (1 = Yes, 0 = No) of rice households, and E_1, E_2, E_3, E_4 are the variables of serious diseases happening during the study year (1 = Yes, 0 = No), rice monoculture (1 = yes, 0 = No), soil quality (1 = fertile soil, 0 = other soils), off-farm income ratio.

Some restrictions are used to check the constant returns to scale:

$$\begin{aligned} \alpha_1 + \alpha_2 + \alpha_3 &= 1 \\ \alpha_{11} + \alpha_{12} + \alpha_{13} &= 0 \\ \alpha_{12} + \alpha_{22} + \alpha_{23} &= 0 \\ \alpha_{13} + \alpha_{23} + \alpha_{33} &= 0 \end{aligned} \tag{7}$$

Then, the following restriction is applied to test the existence of Cobb-Douglass function:

$$\alpha_{11} = \alpha_{12} = \alpha_{13} = \alpha_{22} = \alpha_{23} = \alpha_{33} = 0 \tag{8}$$

3.2. Replacement Cost (RC)

Replacement cost approach is defined as payment for restoring original environment (unpolluted state) if it has already been damaged. The costs of moving away from the polluted area suffered by the victims of environmental damage or actual spending on safeguards against environmental risks are called replacement costs (Bateman et $al.$, 2003; Winpenny, 1991). In the study written by Reddy and Behera (2006), the replacement cost method is used to estimate the damage costs of pump sets due to water pollution. In this study, farmers in the polluted areas might spend more input costs for the compensation of rice productivity loss because they directly use the polluted water for irrigation. Thus, it is assumed that the costs of farmers in polluted areas are more than those in the non-polluted areas. In this case, the replacement cost is estimated by using the cost function approach. The basic form of cost function is given by:

$$C = C(W_s, W_h, W_f, W_p, Y, Z, E, F) \tag{9}$$

where C is the total cost of a farmer (VND/ha), W_s is the price of seed (VND/kg), W_h is the price of herbicide (VND/100ml), W_f is the price of fertilizers (VND/kg), W_p is the price of pesticides (VND/100ml), Y is the rice yield of a farmer in the studied year (tones/ha), Z is a vector of social-economic characteristics of farmers, and E is a vector of farming conditions, F is the relative location of farms (polluted site = 1, non-polluted site = 0)

The test for the existence of cost increase due to water pollution is:

$$\begin{aligned} H_0 &: \text{Cost increase} = 0 \text{ or Coef. of F} = 0 \\ H_1 &: \text{Cost increase} > 0 \text{ or Coef. of F} > 0 \end{aligned} \tag{10}$$

The increase in input costs is defined as the difference of the average cost between heavily polluted and less polluted areas. It is estimated by following equation:

$$\Delta C = C(\overline{W}_s, \overline{W}_h, \overline{W}_f, \overline{W}_p, \overline{Y}, \overline{Z}, \overline{E}, F = 1) - C(\overline{W}_s, \overline{W}_h, \overline{W}_f, \overline{W}_p, \overline{Y}, \overline{Z}, \overline{E}, F = 0) \tag{11}$$

where ΔC is the increase of the average cost per ha because of water pollution (VND/ha); $\overline{W}_s$, $\overline{W}_h$, $\overline{W}_f$, $\overline{W}_p$, $\overline{Y}$, $\overline{Z}$, $\overline{E}$ are the average price of seed, herbicide, fertilizer, pesticides, social-economic characteristics, and farming conditions, respectively.

The Cobb-Douglas formal function is applied to estimate the cost function in the study (Tim & Battese, 2005):

$$\ln(C) = \varphi_0 + \varphi_1 \ln(W_s) + \varphi_2 \ln(W_h) + \varphi_3 \ln(W_f) + \varphi_4 \ln(W_p) + \varphi_5 \ln(Y) + \sum_{k=1}^{3} \beta_k Z_k + \sum_{h=1}^{3} \delta_h E_h + \gamma F \tag{12}$$

where C, W_s, W_h, W_f, W_p, F are the same as in the above equation and Z_1, Z_2, Z_3, are, the age (years), the number of school year (years), attending trainings (1 = Yes, 0 = No) of rice households, and E_1, E_2, E_3, are serious diseases happening during the year (1 = Yes, 0 = No), rice monoculture (1 = yes, 0 = No), soil quality (1 = fertile soil, 0 = other soils) respectively.

3.3. Profit function approach

Net economic return is defined as revenues from rice minus the cost of producing rice. It will be identified by a profit function approach. The profit loss is estimated by the following basic profit function:

$$\pi^* = \pi\left(W^*, C, Z, E, F\right) \tag{13}$$

where π^* is normalized profit defined as gross revenue minus variable cost divided by farm-specific output price, W^* is a vector of variable input prices divided by output price, C is a vector of fixed factors of the farm, Z is a vector of social-economic characteristics of farmers, E is a vector of farming conditions, F is the relative location of farms (polluted site = 1, non-polluted site = 0).

Hypothesis for the existence of profit loss due to water pollution is:

$$H_0 : \pi_n = \pi_p \text{ or Profit Loss} = 0 \text{ or Coef. of } F = 0$$
$$H_1 : \pi_n > \pi_p \text{ or Profit Loss} > 0 \text{ or Coef. of } F < 0 \tag{14}$$

The profit loss due to water pollution is defined by the difference in profit between the polluted and non-polluted areas. It is estimated by the equation:

$$\Delta\pi^* = \pi\left(\overline{W}^*, \overline{C}, \overline{Z}, \overline{E}, F = 0\right) - \pi\left(\overline{W}^*, \overline{C}, \overline{Z}, \overline{E}, F = 1\right) \tag{15}$$

where $\Delta\pi^*$ is Profit loss in 1000 VND/ha. $\overline{W}^*$, $\overline{C}$, $\overline{Z}$, $\overline{E}$ are the average prices of inputs, the average of the fixed factors, the social-economic characteristics of farmers, the farming conditions, respectively.

We use the translog profit functional form. The formula is given as (Rahman, 2002, Surjit & Carlos, 1981)

$$\ln\pi^* = \alpha_0 + \sum_{j=1}^{4}\alpha_j \ln W_j^* + \frac{1}{2}\sum_{j=1}^{4}\sum_{k=1}^{4}\tau_{jk} \ln W_j^* \ln W_k^* + \sum_{j=1}^{4}\sum_{l=1}^{6}\phi_{jl} \ln W_j^* \ln C_l +$$
$$+ \sum_{l=1}^{6}\beta_l \ln C_l + \frac{1}{2}\sum_{l=1}^{6}\sum_{t=1}^{6}\varphi_{lt} \ln C_l \ln C_t + \sum_{m=1}^{3}\varpi_m Z_m + \sum_{n=1}^{4}\eta_n E_n + \gamma F \tag{16}$$

where π^* is the restricted profit (total revenue minus total cost of variable inputs) normalized by price of output (P); W_j^* is the price of the jth input (W_j) normalized by the output price (P); j is the price of seed (1), the price of herbicides (2), the price of fertilizer (3), the price of pesticide (4); C_l is the quantity of fixed input, where l is total amount of seed used (1), total amount of herbicides used (2), total amount of fertilizer used (3), total amount of pesticides used (4), the number of man-days for rice production (5), the money of machines and services at all stages of rice production (6); Z_1, Z_2, Z_3 are the age (years), the number of school year (years), and attendance at training sessions (1 = Yes, 0 = No) of rice households, respectively; and E_1, E_2, E_3, E_4 are the variables of serious disease incidence happening during the study year (1 = Yes, 0 = No), rice monoculture (1 = Yes, 0 = No), soil quality (1 = fertile soil, 0 = other soils), and off-farm income ratio, respectively.

Then, the following restriction is applied to test the existence of the Cobb-Douglass function:

$$\tau_{jk} = \phi_{jl} = \varphi_{lt} = 0 \tag{17}$$

4. Study site and data description

4.1. Study site

In the Mekong River Delta, there are approximately 33 industrial parks, which constitute 9.5% of the total industrial parks of the country. Almost all of these 33 parks have no wastewater treatment system. The industrial parks in Can Tho city have released the biggest pollution loads, and the province is ranked in the top 10 most polluted provinces in Vietnam (Table 1). Can Tho is also one of the biggest rice producers in the Mekong River Delta. Because of these reasons, Can Tho was selected as the study site.

Province	Air index	Land index	Water index	Overall
Ho Chi Minh city	1	1	1	1
Hanoi	5	2	2	2
HaiPhong	2	6	4	3
Binh Duong	6	3	3	4
Dong Nai	4	4	5	5
Thai Nguyen	3	5	7	6
PhuTho	7	7	6	7
Da Nang	10	9	8	8
Ba RiaVung Tau	9	8	10	9
Can Tho	8	10	9	10

Note: The pollution loads released to air, land and water were estimated for all 64 provinces in Vietnam, and then pollution indexes were calculated and rankings were made.

Source: ICEM, 2007

Table 1. Top 10 most polluted provinces in Vietnam

Zones	Size	Main activities	Water treatments
Tra Noc 1	135 ha	Processing, electron, clothes	No [a]
Tra Noc 2	165 ha	Machinery	No [a]
Hung Phu 1	262 ha	Harbor, Store	No
Hung Phu 2	212 ha	Machinery	No
Hong Bang	38.2 ha	Consumer goods	No [a]
Thot Not	150 ha	Processing, clothes, shoes	No [a]

[a] The available decision and acceptation of local authorities to evaluate the impact of environmental pollution.

Source: Resource and Environment Department of Can Tho City (2008)

Table 2. The industrial zones in Can Tho city

There are six industrial parks in Can Tho (Table 2), which mainly comprise agricultural and fishery processing industries, clothes and consumer goods manufacturing industries. Almost none of the industrial zones and industrial corporations located near human residences have installed wastewater treatment systems. There has been little management of toxic waste or water pollution by local authorities and business. Tra Noc 1 (built in 1995) and Tra Noc 2 (built in 1999) industrial zones have only recently been acknowledged by the Department of Resources and Environment while Thot Not has been considered by Can Tho authorities to evaluate the impact of environmental pollution (Resource and Environment department of Can Tho city, 2008). As a consequence, Tra Noc 1 and 2 have released large volumes (1000s m^3) of various waste products directly into the river (Tuyen, 2010).

4.2. Data collection

The study region covers the area within and around Tra Noc 1 and Tra Noc 2 industrial zones, which are two of the greatest polluters in Can Tho. People living in this area have suffered various financial impacts from the pollution: reduced crop yields, the use of cattle and agricultural equipment such as pump sets, contamination of drinking water, and increased incidence of human diseases and deaths directly and indirectly caused by water pollution.

Farmers were randomly selected for interview from two areas (Phuoc Thoi and Thoi An) with similar social and natural conditions (e.g. the same social and farming culture, ethnicity, type of soil). The selection of the polluted and non-polluted area was based on their distance from industrial zones, and on the recommendation or suggestion of local authorities and farmers. Some of the villages in Phuoc Thoi are heavily polluted by wastewater from the TraNoc 1 and 2 industrial zones. The villages in Thoi An are further away from the industrial zones than Phuoc Thoi and deemed to represent a non-polluted area (see Figure 1).

The group of fourteen interviewers and three local guide persons includes ten final year students, four staffs of School of Economics and Business Administration, Can Tho University, one local authority from people's committee, and two local farmers.

The questionnaire composes four main parts. In the first and second parts, the personal and farming information of household such as address, age, gender, training and so on and the situation of environmental pollution were interviewed. The inputs and output of rice production were collected in the three part and income from other activities obtained in the final section of questionnaire.

The household survey took 3 months to complete from January to March 2010 and was divided into two main reporting periods. The first period was called as pilot-survey in January 2010. The aims of this interview were to check and then correct the questionnaire more clearly and concisely, and to help interviewers get used to and understand the content of questionnaire. After the interviewers were trained how to ask by using questionnaire, about 30 farmers were interviewed. The revised questionnaire was used in the second period from February to March 2010. In total, 364 rice farmers, consisting of 214 farmers in the polluted and 150 farmers in the non-polluted area, were interviewed in February and March 2010. Household data were collected on household level information related to production costs and income as well as the social and economic characteristics of the farmers, and their perceived damages and losses due to water pollution.

Table 3 showed the water quality index of the polluted and non-polluted area. The concentrations of Total Suspended Solids (TSS) in the water refer to the concentrations of solid particles that can be trapped by a filter. This can be a problem because high concentrations of TSS can block sunlight from reaching submerged vegetation. This causes a reduction in the photosynthesis rate, and therefore less dissolved oxygen released into the water by plants. If bottom dwelling plants are not exposed to some light, the plants stop producing oxygen and die. Chemical Oxygen Demand (COD) is the amount of oxygen used during the oxidation of organic matter and inorganic chemicals such as ammonia nitrogen (NH_3-N). High COD indicates a greater pollution load.

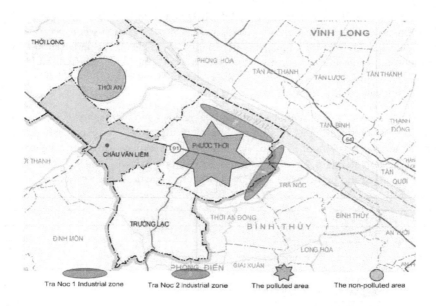

Figure 1. Map of the Study Site

	TSS (mg/l)	COD (mg/l)	NH3-N (mg/l)
The polluted area (PhuocThoi) [1]			
- Sewer mouth	145	720	13.29
- Primary affected water source [2]	50	50	1.23
- Secondary affected water source [3]	60	48	0.63
The non-polluted area (Thoi An)[4]	22	5.1	0.16
Limitation value (TCVN5942,1995)			
- Class A [5]	20	10	0.05
- Class B [6]	80	35	1

Notes:

(1) Measured on January 17th, 2007 (Nga et al., 2008)

(2) The region receives wastewater directly from the industrial park.

(3) The region receives polluted water from the primary affected water source regions.

(4) Measured on January 27th, 2007 (Lang et al., 2009)

(5) Values in Class A are from the surface water used for domestic water supply with appropriate treatments.

(6) Values in Class B are from the surface water used for purposes other than domestic water supply. Water quality criteria for aquatic life are specified in a separate index.

Table 3. Water quality of the polluted and non-polluted area

In the polluted area, the concentrations of TSS, COD and NH3-H in the sewer mouth, the primary affected water source and the secondary affected water source regions were mostly much higher than those of the standard water quality (see Table 3). This indicated that our selected pollution area site was heavily polluted. The concentrations of TSS, COD and NH3-N in the sewer mouth region were nearly 2-fold, over 20-fold and 13-fold higher than those of the standard water quality of class B, respectively.

Differences in the water quality index between the polluted and non-polluted area indicate that the water quality in the non-polluted area was much higher than that in the polluted area. However, the concentrations of TSS and NH3-N in the non-polluted were slightly higher than those of the Class A standard. This may be caused by non-point source pollutants, for instance, fertilizer, herbicide and pesticide released by agricultural activities in the region.

Variable	Description	Unit
Y	Total yield per hectare	Ton/hectare
P	Price of rice	Thousand VND/ton
C	Total cost	Thousand VND/ha
π	Total profit	Thousand VND/ha
C_s	Total amount of seed used	Kg/ha
C_h	Total amount of herbicides used	Equivalent unit of 100 ml/ha
C_f	Total amount of fertilizer used	Kg/ha
C_p	Total amount of pesticide used	Equivalent unit of 100 ml/ha
C_l	The number of man-days for rice production	day/ha
C_c	The money of machines and services at all stages of rice production	Thousand VND/ha
W_s	Price of seed	Thousand VND/kg
W_h	Price of herbicide	Thousand VND/100ml
W_f	Price of fertilizer	Thousand VND/kg
W_p	Price of pesticide	Thousand VND/100ml
Age	The age of respondents	Years
Education	The number of school year of respondents	Years
Training	Respondents attending trainings	1= Yes, 0 = No
Mono	Rice monoculture	1= Yes, 0 = No
Diseases	Diseases happening during the study year	1= Yes, 0 = No
Off-farm ratio	The ratio of off-farm income	
Soil	Soil quality	1 = fertile soil, 0 = other soils

Table 4. Description of variables used in rice production models

Table 4 showed the descriptions of variables in rice production models. The volumes of herbicide and pesticide used have measurement units of equivalent units of 100 ml per hectare per crop, based on farmers' reports and experts' recommendations. This is because farmers use various types of herbicides and pesticides (mixed with water or as a powder), and sometimes mix them together, which means that it is difficult to estimate exact amounts.

Variables	Non-polluted area	Polluted area	t-value
Y	5.88	4.99	-7.31***
P	4,157.79	4,060.89	-3.06***
C	10,909.44	10,563.61	-0.84
π	13,623.91	9,759.35	-8.37***
C_s	224.41	206.42	-2.37**
C_h	10.43	11.70	1.33
C_f	475.87	463.76	-0.53
C_p	77.26	70.23	-1.24
C_i	29.03	32.95	1.39
C_c	3283.02	3436.20	1.06
W_s	5.46	5.21	-1.34
W_h	32.79	32.47	0.21
W_f	9.42	9.54	0.92
W_p	24.86	21.70	-1.85*
Age	48.04	48.99	0.81
Education	6.33	6.07	-0.87
Training	0.49	0.35	-2.72***ψ
Mono	0.60	0.58	-0.39ψ
Diseases	0.40	0.42	-0.39ψ
Off-farm ratio	0.20	0.37	4.72***ψ
Soil	0.63	0.75	2.35**ψ

Notes: ***, **, * indicate statistical significance at the 0.01, 0.05 and 0.1 level respectively

ψ Z-test for the equality of two proportions

Source: Own estimates; data appendix available from authors.

Table 5. Descriptive Statistics of Rice Production per hectare per crop

Table 5 showed the descriptive statistics of the main variables in the rice production model for the polluted and non-polluted areas. Although soil quality in the non-polluted area was significantly ($P < 0.05$) lower than that in the polluted area, rice productivity and profit in the non-polluted area was significantly ($P < 0.01$) higher than those in the polluted area. The price of rice in the polluted area was significantly ($P < 0.01$) lower than that in the non-polluted area. This indicated that water pollution might have reduced crop quality, and in turn its price. The difference in the off-farm income ratio between the two areas suggests that farmers are aware of the reduced profit from rice cultivation in polluted soil, and therefore have a tendency to find additional work in nearby industrial parks to supplement their income.

Other variables measured did not significantly differ between the two regions (Table 5), except the percentage of respondents attending training. The results also showed that, on average, farmers were 48 years old, have had 6 years of education and 60 % of them grew rice in a monoculture.

5. Estimated results

5.1. Impact of water pollution on rice productivity

Table 6 showed the Ordinary Least Squares (OLS) result of rice production function in translog form. The variables estimated in the model were statistically significant at 1 percent level. The estimated R-square was equal to 0.64, revealing the 64 percent change of rice yield possibly explained by independent variables in the model.

Variables	Coef.	t-value	Variables	Coef.	t-value
$\ln(C_s)$	0.696	0.87	$\ln(C_h)\times\ln(C_p)$	0.007	0.7
$\ln(C_h)$	-0.123	-0.8	$\ln(C_h)\times\ln(C_l)$	0.021	1.49
$\ln(C_t)$	0.465	0.84	$\ln(C_h)\times\ln(C_c)$	0.008	0.35
$\ln(C_p)$	-0.157	-0.59	$\ln(C_t)\times\ln(C_p)$	0.060	1.5
$\ln(C_l)$	0.851**	2.55	$\ln(C_t)\times\ln(C_l)$	-0.026	-0.56
$\ln(C_c)$	0.572	1.3	$\ln(C_t)\times\ln(C_c)$	-0.022	-0.27
½ $\ln(C_s)^2$	0.532***	2.72	$\ln(C_p)\times\ln(C_l)$	-0.031	-1.41
½ $\ln(C_h)^2$	0.000	0.03	$\ln(C_p)\times\ln(C_c)$	0.029	0.71
½ $\ln(C_t)^2$	0.037	1.13	$\ln(C_l)\times\ln(C_c)$	-0.023	-0.58
½ $\ln(C_p)^2$	-0.011	-0.66	Age	-0.002**	-2.07
½ $\ln(C_l)^2$	0.059*	1.81	Education	0.004	1.1
½ $\ln(C_c)^2$	0.075	1.07	Training	0.039**	2.07
$\ln(C_s)\times\ln(C_h)$	0.014	0.54	Disease	-0.012	-0.67
$\ln(C_s)\times\ln(C_t)$	-0.132	-1.34	Mono	0.016	0.7
$\ln(C_s)\times\ln(C_p)$	-0.057	-1.08	Soil	0.031	1.64
$\ln(C_s)\times\ln(C_l))$	-0.105*	-1.83	Off-farm ratio	-0.054*	-1.93
$\ln(C_s)\times\ln(C_c)$	-0.216*	-1.87	Pollution	-0.127***	-6.68
$\ln(C_h)\times\ln(C_t)$	-0.025	-0.7	Constant	-5.615**	-2.26
R-square					0.64
Included observation					364

Notes: ***, **, * indicate statistical significance at the 0.01, 0.05 and 0.1 level respectively

Source: Own estimates; data appendix available from authors.

Table 6. The OLS regression of rice production function

The study also examined the null hypothesis in (7) that there was a proportional output change when inputs in the model were varied or farms produce rice with constant returns to scale. The restricted least squares regression with the null hypothesis of constant returns to scale was estimated. The computed F statistic was 37.09 more than the critical value F (7, 327) of 2.69 at

1 percent level of significance [1]. Thus, the null hypothesis was rejected and the study concluded that technology did not exhibit constant returns to scale.

The second test was applied to check the Cobb-Douglass formal existence of the production function. The restricted function was estimated with the null hypothesis of jointed parameters in (8) equal to 0. The computed F statistic of 1.94 was more than the critical F(21,327) of 1.91 at 1 percent level of significance [1]. Thus, the null hypothesis was rejected, meaning that the translog functional form was suitably applied for the data of rice production in the study.

The results of Table 6 showed that there was no multicollinearity in the independent variables of production function because the correlations of these independent variables estimated by using the correlation matrix were less than 70 percent. The null hypothesis homoscedasticity was also accepted by using Breusch-Pagan test because the estimated LM of 49.72 was less than the critical χ^2_{36} of 57.34 at the level of 1 percent [2].

Table 6 showed that the rice productivity in the polluted was lower than in the non-polluted area because the coefficient of *Pollution* variable was significantly negative at 1 percent level. In addition, the study also revealed that training courses partly contributed an increase in rice yield since the coefficient of *Training* variable was significantly positive.

Moreover, the model also showed that farmer age (P < 0.05) and the ratio of off-farm income (P < 0.1) explained variation in rice yield. The effect of age might have been caused by declines in the health of older farmers leading to less efficient cultivation. Farmers who earned more off-farm income were associated with less profitable rice cultivation. Our interviews with the farmers in the polluted region suggested that when rice production was no longer profitable, farmers tended to sell their land as construction land or rent their land to farmers from other regions. Local farmers also attempted to secure employment in the nearby industrial parks, from which they could earn more money than compared to rice cultivation. The study also discovered that water pollution made farmers change rice cultivation and crop intensification techniques. Before their income was mainly from rice production with three rice crops per year, now they do rice farming as part-time jobs, only grow one or two crops per year and harvest rice just enough for home consumption. These possibly were the suitable explanations for the negative impact of off-farm income on rice productivity.

The reduced productivity of rice was calculated based on findings from Table 6. After the equation (5) was used to eliminate the effects of other factors, the estimated yield in the non-polluted area was about 5.61 tons and around 4.94 tons for the polluted region. Then, the loss of rice yield due to polluted water irrigation was estimated by subtracting the yield in the polluted from yield in the non-polluted region (equation 5). Using this approach, the estimated result was about 0.67 tons per hectare per crop (5.61 tons – 4.94 tons).

5.2. Increase in rice production cost due to water pollution

Table 7 showed R-square was equal to 0.56, revealing the variation of total rice costs of 56 percent was explained by independent variables in the model. The study also showed that the multicollinearity among the independent variables in cost function did not exist because the results estimated by correlation matrix approach showed that there were no correlations in these independent variables higher than 70 percent. The result of Breusch-Pagan test performed that the estimated LM of 14.96 was less than the critical χ_{12}^2 of 26.22 at the level of 1 percent, revealing the absence of heterscedasticity in the estimate of cost function [2].

Variables	Coefficient	t-value
ln(W_s)	0.195***	3.95
ln(W_h)	0.021	0.84
ln(W_t)	0.431***	5.5
ln(W_p)	0.007	0.27
ln(Y)	0.918***	14.79
Age	0.002*	1.9
Education	-0.008	-1.55
Training	-0.058**	-2.02
Diseases	0.045	1.62
Rice monoculture	0.143***	4.72
Soil	-0.032	-1.08
Pollution	0.098***	3.3
Constant	6.140***	22.84
Statistic summary		
R-square	0.56	
Included observation	364	

Notes: ***, **, * indicate statistical significance at the 0.01, 0.05 and 0.1 level respectively

Source: Own estimates; data appendix available from authors.

Table 7. The OLS regression of rice cost function

The coefficient of *Pollution* variable was statistically significant positive at level of 1 percent, performing rice costs in the polluted region was higher than one in the non-polluted region. Moreover, farmers, who were older, managed their production cost more highly and less efficiently, performed by the positive effect of *Age* variable on total costs at 10 percent level. The significantly positive coefficient of *Rice monoculture* variable (P < 0.01) revealed that farmers who grew rice monoculture cost more than ones who cultivated rice rotation or intercropping. Possible explanation is that the cropping system of rice monoculture decreased the fertility of soil.

Like the calculation of yield loss, cost increase due to water pollution was estimated using the coefficients performed in Table 7. After the effect of other factors were eliminated, total cost was estimated about VND 10.37 million for rice production in the polluted area and VND 9.4

per ha per crop for that in the non-polluted area. Cost increase was estimated by subtracting the rice cost in the non-polluted region by the rice cost in the polluted area (equation 11). Using this approach, an increase in cost due to water pollution was calculated around VND 0.97 million per ha per crop (See Table 10).

5.3. Total loss of net economic return

Table 8 showed the coefficients from the OLS regression of the rice profit model using the translog profit functional form (equation 16). The full model was statistically significant at the 1% level. The estimated R-square revealed that 50% of the variation in the rice profit was explained by the model.

Next, we tested the null hypothesis of the Cobb-Douglass functional form. The restricted function was estimated assuming the null hypothesis that the joint parameters in (17) are 0. The computed F statistic of 1.78 was more than the critical F(55,283) value of 1.57 at the 1 percent level [1]. The null hypothesis was therefore rejected, which supported the use of the translog functional form in this study. The estimate of profit function also showed the absence of multicollinearity (the correlations of independent variables less than 70 percent) and of heterscedasticity (Breusch-Pagan test showed the critical χ_{74}^2 of 105.2 at the level of 1 percent higher than the computed LM of 100.24) [2].

The coefficient of *Pollution* variable representing the effect of pollution was negative and significant (P < 0.01), which confirmed that water pollution reduced the profit of rice cultivation. The reduction in rice profit was calculated using the coefficients presented in Table 8. The estimated profit was approximately VND 9.14 million for rice cultivation in the polluted area and VND 12.34 million for that in the non-polluted area after the influences of other factors were eliminated. The loss of rice profit due to wastewater irrigation was estimated by subtracting the rice profit in the polluted region by the rice profit in the non-polluted region (equation 15). Using this approach, the loss of profit was calculated to be approximately VND 3.2 million per hectare per crop (see Table 10).

Like the results of rice yield loss, this model also performed that farmer age (P < 0.01), attending training (P < 0.01) and the ratio of off-farm income (P < 0.1) explained variation in profit. Moreover, soil quality was also an important factor affecting profit (P<0.1).

We also used the same estimate of profit loss due to water pollution to calculate reductions in profit caused by other factors as presented in Table 9. Cultivation in non-fertile soil, instead of fertile soil, could reduce rice profit by 8.24%. Farmers whose main sources of income were from non-agricultural sectors obtained 11.45% less rice profit than those who only had an agricultural income. Participating in trainings was estimated to increase profit by 13.03%. Profit loss caused by water pollution was much higher than the profit loss caused by other factors, which demonstrates that environment pollution has a great significance for rice farmers near industrial parks. Because of this, we suggest that the Vietnamese authorities should place a greater importance on the development and implementation of pollution control policies.

Variables	Coef.	t-value	Variables	Coef.	t-value
$\ln(W_s^*)$	-0.111	-0.02	$\ln(W_p^*)\times\ln(C_c)$	0.162	0.92
$\ln(W_h^*)$	1.211	0.54	$\ln(C_s)$	-3.393	-0.72
$\ln(W_f^*)$	-3.869	-0.71	$\ln(C_h)$	-0.540	-0.34
$\ln(W_p^*)$	-0.698	-0.29	$\ln(C_f)$	3.341	0.99
$\frac{1}{2}\ln(W_s^*)^2$	0.322	0.68	$\ln(C_p)$	-2.241	-1.16
$\frac{1}{2}\ln(W_h^*)^2$	-0.019	-0.15	$\ln(C_i)$	2.890	1.51
$\frac{1}{2}\ln(W_f^*)^2$	-1.245**	-1.99	$\ln(C_c)$	1.432	0.37
$\frac{1}{2}\ln(W_p^*)^2$	0.223	1.58	$\frac{1}{2}\ln(C_s)^2$	0.338	0.61
$\ln(W_s^*)\times\ln(W_h^*)$	-0.017	-0.09	$\frac{1}{2}\ln(C_h)^2$	0.012	0.18
$\ln(W_s^*)\times\ln(W_f^*)$	-0.257	-0.50	$\frac{1}{2}\ln(C_f)^2$	-0.117	-0.47
$\ln(W_s^*)\times\ln(W_p^*)$	-0.042	-0.22	$\frac{1}{2}\ln(C_p)^2$	0.013	0.18
$\ln(W_h^*)\times\ln(W_f^*)$	0.231	0.91	$\frac{1}{2}\ln(C_i)^2$	0.123	1.44
$\ln(W_h^*)\times\ln(W_p^*)$	0.030	0.28	$\frac{1}{2}\ln(C_c)^2$	-0.107	-0.52
$\ln(W_f^*)\times\ln(W_p^*)$	0.005	0.02	$\ln(C_s)\times\ln(C_h)$	-0.113	-0.80
$\ln(W_s^*)\times\ln(C_s)$	-0.648*	-1.96	$\ln(C_s)\times\ln(C_f)$	0.500*	1.73
$\ln(W_s^*)\times\ln(C_h)$	0.181	1.32	$\ln(C_s)\times\ln(C_p)$	0.013	0.06
$\ln(W_s^*)\times\ln(C_f)$	0.586**	2.01	$\ln(C_s)\times\ln(C_i))$	-0.250	-1.56
$\ln(W_s^*)\times\ln(C_p)$	-0.165	-0.96	$\ln(C_s)\times\ln(C_c)$	-0.149	-0.48
$\ln(W_s^*)\times\ln(C_i)$	0.079	0.56	$\ln(C_h)\times\ln(C_f)$	-0.130	-1.19
$\ln(W_s^*)\times\ln(C_c)$	0.029	0.08	$\ln(C_h)\times\ln(C_p)$	0.054	0.93
$\ln(W_h^*)\times\ln(C_s)$	-0.314	-1.62	$\ln(C_h)\times\ln(C_i)$	0.148**	2.33
$\ln(W_h^*)\times\ln(C_h)$	-0.064	-0.80	$\ln(C_h)\times\ln(C_c)$	0.012	0.10
$\ln(W_h^*)\times\ln(C_f)$	-0.124	-0.84	$\ln(C_f)\times\ln(C_p)$	-0.137	-1.08
$\ln(W_h^*)\times\ln(C_p)$	0.087	0.89	$\ln(C_f)\times\ln(C_i)$	-0.055	-0.41
$\ln(W_h^*)\times\ln(C_i)$	0.115	1.41	$\ln(C_f)\times\ln(C_c)$	-0.568**	-2.46
$\ln(W_h^*)\times\ln(C_c)$	0.244	1.63	$\ln(C_p)\times\ln(C_i)$	-0.140**	-2.07
$\ln(W_f^*)\times\ln(C_s)$	0.533	1.01	$\ln(C_p)\times\ln(C_c)$	0.192	1.45
$\ln(W_f^*)\times\ln(C_h)$	-0.376*	-1.83	$\ln(C_i)\times\ln(C_c)$	-0.025	-0.22
$\ln(W_f^*)\times\ln(C_f)$	-0.609	-1.63	Age	-0.006***	-2.63
$\ln(W_f^*)\times\ln(C_p)$	-0.104	-0.45	Education	0.010	1.22
$\ln(W_f^*)\times\ln(C_i)$	0.106	0.47	Training	0.140***	2.90
$\ln(W_f^*)\times\ln(C_c)$	-0.529	-1.26	Disease	0.016	0.35
$\ln(W_p^*)\times\ln(C_s)$	0.025	0.12	Mono	0.003	0.06

Variables	Coef.	t-value	Variables	Coef.	t-value
$\ln(W_p{}^*)\times\ln(C_h)$	-0.010	-0.15	*Soil*	0.086*	1.75
$\ln(W_p{}^*)\times\ln(C_e)$	-0.164	-1.11	*Off-farm ratio*	-0.126*	-1.73
$\ln(W_p{}^*)\times\ln(C_p)$	0.107	1.14	*Pollution*	-0.300***	-5.81
$\ln(W_p{}^*)\times\ln(C_l)$	-0.098	-1.22	Constant	-20.213	-0.60
R-square			0.50		
Included observation			364		

Table 8. The OLS regression of rice profit function

Factors	Reduced profit (Thousand VND)	Percentage of reduced profit (%)
Polluted vs. Non-polluted area	3,203	25.95
Non-fertile vs. Fertile soil	874	8.24
Non-training vs. Training	1,465	13.03
The highest off-farm vs. Zero off-farm income ratio	1,229	11.45

Source: Own estimates; data appendix available from authors.

Table 9. Reduced profit in rice farming and key constraints

Table 10 summarized the total loss of rice production due to water pollution. The estimated results showed there were about 26 percent of profit loss, including around 12 percent of reduced quantity (yield loss) and 9 percent of cost increase, adversely caused by industrial water pollution. In this study, we also observed that farmers in the polluted area use water irrigation from the highest water tide level to reduce the effects of wastewater on rice production. This was because the farmers thought the water at the high tide level looked less polluted than the waters at other times, despite the fact that the water was always heavily polluted near the industrial parks.

	Amount	Percent
Quantity loss	0.67 tons/ha	12%
Cost increase	0.97 million VND/ha	9%
Total loss of net economic return	3.2 million VND/ha	26%

Source: Own estimates; data appendix available from authors.

Table 10. Impact of water pollution on rice production

Moreover, the use of polluted water also caused the farmers to change their cultivation management. In previous years, three rice crops were produced annually and rice cultivation was the main income source. However, because of pollution, only one or two rice crops is now cultivated in the polluted area each year, and farmers treat rice cultivation as a part-time job, producing rice sufficient only for household consumption.

During our study, we also received reports of skin diseases on the farmers working in the polluted region. For instance, a farmer in the polluted area reported that he had suffered from skin disease 5 days per year, and the treatment cost VND 500,000. The diseases also caused the loss of 2.5 workdays, equivalent to VND 250,000. Therefore, the estimate of total economic loss is underestimated if indirect costs such as the health costs suffered by farmers are not included.

6. Conclusions and policy implication

Local authorities in Vietnam have recently removed or reduced some of the environmental impact requirements to attract industrial investments to their province. Although industrial investments with low environmental standards might increase gross domestic product and create more jobs for local households, they may also bring many problems including water, air and soil pollution. This study provides an example of the negative impacts that arise from pollution by industries.

In this study, we surveyed rice farmers in two areas with the same natural environment conditions, social characteristics (e.g. the same social and farming culture, ethnicity, type of soil), and only differed with respect to pollution. One area was considered to be the polluted area, receiving wastewater from nearby industrial parks, while the other area was assumed to be the non-polluted area, being distant from sources of industrial pollutants. The productivity loss of rice production caused by water pollution was estimated by the difference in rice yield between the two regions. The similar calculation was applied for cost increase and profit loss for using wastewater irrigation. The results showed that the yield loss of rice was about 0.67 tons per hectare per crop, VND 0.97 million for cost increase and totally 26 percent of profit loss due to water pollution. Therefore, since the study includes 214 farmers in the polluted area and these 214 farmers cultivate rice in 148 hectare as a whole, their total cost increase per crop because of water pollution could be estimated about VND 144 million (VND 0.97 * 148ha) and approximately VND 474 million (VND 3.2 million * 148ha) for their total net economic loss.

According to The World Bank (2007), the development of rice roots and seedlings could be influenced by using wastewater for irrigation. Polluted water irrigation causes the reduction of height, leaf area and dry matter. Decrease in leaf surface area leads to the reduction of photosynthesis. These facts have directly impact on rice production. In other words, the impacts of polluted water on rice productivity mainly reduce the number of ears unit area, number of seed per ear and seed weight. The study estimated water pollution caused yield reduction about 12 percent. This result is nearly equal to the reduced yield of 10 percent in the sewage-irrigated area in comparison with clear water-irrigated areas estimated by Bai (2004), but much lower than the rice reduced productivity of 20 percent calculated by Song (2004) in the study of Lindhjem (2007) and 30 percent by Chang *et al.* (2001).

Economic developments that cause damage to natural resources and the environment are unsustainable. We suggest that the Vietnamese government needs to develop policies that ensure sustainable development. Similar to environmental policies in developed countries, the Vietnamese government could consider increasing the current environmental standards and raising environmental taxes. The increase of environmental taxes could not only encourage industries to apply new technologies that reduce environmental pollution, but also generate money to compensate farmers near industrial areas for the damage to their agricultural production and health and to build wastewater treatment facilities in industrial parks. Compensation could be provided directly in cash to the farmers, or indirectly by means such as funding training or activities related to new technologies and the management of agricultural inputs and expenditure. Our study showed that training helped farmers increase their profit, which might partly offset some of the losses caused by environmental pollution.

To reduce polluted water from the industrial parks, an increase in the effectiveness of implementation of Decision 64 and Circular 07 should be recommended. A public disclosure system for the environmental performance of polluters mentioned in Article 104 of the Law on Environmental Protection (dated 2005) and Article 23 of Degree No. 80/2006ND-CP should be considered as one of the best ways to increase the efficiency of Decision 64 and Circular 07.

Article 104 requires polluters to report and publicize the information and data about the environment as follows:

- Reports on the environmental impact assessment, decision on approval for reports on the environmental impact assessment and plan for the implementation of requirements stipulated in the decision on approval for reports on the environmental impact assessment;

- List of and information about sources of wastes, pollutants that seem potentially harmful to people's health and environment;

- Areas where environment is polluted and degraded seriously and extremely seriously, areas in danger of the environmental pollution.

- Report on the environmental situation at the provincial level, report on environmental impact assessment by industries, fields and the national report on the environment

- It is essential to ensure unrestricted access to publicized information

- Agencies publicizing information about the environment have to take responsibility on accuracy, honesty and objectivity of announced information before legal agencies.

Article 23 provides details and instructions on how to implement Article 104 of the Law on Environmental Protection. These details and instructions include:

- The Ministry of Natural Resources and Environment have responsibility for announcing information and data about the national environment;

- Ministries and ministerial-level agencies, government agencies shoulder responsibility for exposing information and data about the environment in industries and areas under their management;

- Agencies in charge of the environmental protection of People's Committees at all levels bear responsibility for make information and data about the environment in the area under their management publicly;

- Management board of economic zones, industrial parks, export processing zones, managers of manufacturing and service units accept responsibility for publicizing information and data about the environment in the area under their management;

- Publicity of information and data about the environment is stipulated as follows:

- Information and data about the environment is publicized in form of books, news in newspapers or post on units' websites;

- Information and data about the environment is publicized in form of books, news in newspapers or post on units' websites (if any), reported in people's council meetings, announced on notice boards in residential quarter meetings, or listed in headquarters of units or headquarters of commune, ward, town people's committee where units are in operation.

The requirements of these above public disclosure system illustrate a new and significant approach for environmental authorities to force environmental laws and regulations in strong manner by increasing environmental awareness and permitting the large public to put pressure on polluters to solve current environmental problems. Such public disclosure requirements also create significant pressure on environmental authorities themselves as their own decision failures might also be widely recognized by such requirements. However, the implementation of these requirements in a clear, precise, and systematic manner is strongly needed.

Since water treatment facilities in these industrial parks must be built as soon as possible, the study on their cost effectiveness could be needed and seriously considered to decide whether we should build the water treatment facilities in every individual factory or for the whole industrial parks. Moreover, we suggest that the government should not use high-yield agricultural land for the construction of new industrial parks unless they include the latest pollution treatment technologies. The impact of environmental pollution should continue to be evaluated.

Notes

[1] Calculated by the formula $F = \dfrac{(RSS_R - RSS_U)/J}{RSS_U/(N-K)}$, where RSS_R and RSS_U are the restricted and unrestricted sums of squared residuals, J is the number of restrictions, N is the number of observations, and K is the number of parameters in an unrestricted function.

2) Breusch-Pagan test for heterscedasticity:

$$LM = nR^2 \sim X^2_k$$

where: n is the number of observations

R^2 is the R-Square of $|\hat{u}_i| = \delta_0 + \delta_1 X_{1i} + \delta_2 X_{2i} + \ldots + \delta_k X_{ki} + \tilde{v}_i$

k is the number of restricted factors

Acknowledgments

We would like to express our gratitude to Dr. Benoit Laplante, a long-standing resource person with EEPSEA, and Dr Herminia Francisco, a director of EEPSEA, for their invaluable comments and suggestions in shaping the structure of the study and data analyses. Our warmest thanks go to EESEA for sponsoring the study, to Mr. Joseph Arbiol in our laboratory for reading the draft of the manuscript, and to our colleagues at Can Tho University for assisting in data collection.

Author details

Huynh Viet Khai[1] and Mitsuyasu Yabe[2]

1 Can Tho University, Vietnam

2 Kyushu University, Japan

References

[1] Bai, Y. (1988). Pollution of irrigation water and its effects, Beijing, China, Beijing Agriculture University Press: Beijing Agriculture University Press.

[2] Bateman, I. J, Carson, R. T, Day, B, Hanemann, M, Hanley, N, Hett, T, Lee, M. J, Loomes, G, Mourato, S, Ozdemiroglu, E, Pearce, D. W, Sugden, R, & Swanson, J. (2003). Economic Valuation With Stated Preference Techniques Edward Elgar Publishing, London.

[3] Chang, Y, Hans, M. S, & Haakon, V. (2001). The Environmental Cost of Water Pollution in Chongqing, China. Environment and Development Economics, 6(313-333)

[4] Do, T. N, & Bennett, J. Would wet biodiversity conservation improve social welfare? A case study in Vietnam's Mekong River Delta, International Conference on Sustainable Development: Challanges and Opportunities for GMS,

[5] Hung, P. T, Tuan, B. A, & Chinh, N. T. (2008). The Impact of Trade Liberalization on Industrial Pollution: Empirical Evidence from Vietnam the Economy and Environment Program for Southeast Asia (EEPSEA).

[6] ICEM ((2007). Analysis of Pollution from Manufacturing Sectors in Vietnam International Centre for Environmental Management, Indooroopily, Queenland.

[7] Khai, H. V, & Yabe, M. (2011). Evaluation of the Impact of Water Pollution on Rice Production in the Mekong Delta, Vietnam. The International Journal of Environmental, Cultural, Economic and Social Sustainability , 7(5)

[8] Khai, H. V, & Yabe, M. (2012). Rice Yield Loss Due to Industrial Pollution in Vietnam. Journal of US-China Public Administration , 9(3)

[9] Kompas, T. (2004). Market Reform, Productivity and Efficiency in Vietnamese Rice Production. International and Development Economics Working Papers, Asia Pacific School of Economics and Governmnet, Australian National University, Australia.

[10] Lang, V. T, Vinh, K. Q, & Truc, N. T. T. (2009). Environmental Consequences of and Pollution Control Options for Pond "Tra" Fish Production in Thotnot District, Can Tho city, Vietnam The Economy and Environment Program for Southeast Asia (EEPSEA).

[11] Lindhjem, H. (2007). Emvironmental Economic Impact Assessment in China: Problems and Prospects. Environmnetal Impact Assessment Review , 27(1)

[12] Linh, H. V. (2007). Efficiency of Rice Farming Households in Vietnam: A DEA With Bootstrap and Stochastic Frontier Application. University of Minnessota, Minesota, USA.

[13] Nga, B. T, Giao, N. T, & Nu, P. V. (2008). Effects of Waste Water From Tra Noc Industrial Zone on Adjacent Rivers in Can Tho City. Scientific journal of Can Tho University , 9(2)

[14] Quang, M. N. (2001). An Evaluation of the Chemical Pollution in Vietnam.Retrieved May 19, 2011, from http://www.mekonginfo.org/mrc_en/doclib.nsf/ 38bfa13a79f297d2c72566170044aaf9/1d952c500be72dc587256b74000703c8?OpenDocument

[15] Rahman, S. (2002). Profit Efficiency Among Bangladeshi Rice Farmers. Food Policy , 28(5-6)

[16] Reddy, V. R, & Behera, B. (2006). Impact of Water Pollution on Rural Communities: An Economic Analysis. Ecological Economics , 58(3)

[17] Resource and Environment department of Can Tho city ((2008). Report on the situation of Environment in Can Tho City.

[18] Surjit, S. S, & Carlos, A. B. (1981). Estimating Farm-Level Input Demand and Wheat Supply in the Indian Punjab Using a Translog Profit Function. American Journal of Agricultural Economics , 63(2)

[19] The World Bank ((2007). Cost of Pollution in China- Economic Estimates of Physical Damages.

[20] The World Bank ((2008). Handling Serious Environmnetal Polluters in Vietnam: Review of Implementation and Recommendations.

[21] Thong, L. Q, & Ngoc, N. A. (2004). Incentives for Wastewater Management in Industrial Estates in Vietnam the Economy and Environment Program for Southeast Asia (EEPSEA).

[22] Tim, C. D. S. P, & Battese, G. E. (2005). An Introduction to Efficiency and Productivity Analysis, New York, Springer Science.

[23] Tuyen, B. C. (2010). Strategy of Vietnam towards addressing Environment and Climate risks.

Chapter 5

Ultrasonic Membrane Anaerobic System (UMAS) for Palm Oil Mill Effluent (POME) Treatment

N.H. Abdurahman, N.H. Azhari and Y.M. Rosli

Additional information is available at the end of the chapter

1. Introduction

Palm oil mill effluent (POME) is an important source of inland water pollution when released into local rivers or lakes without treatment. The production of palm oil, however, results in the generation of large quantities of polluted wastewater commonly referred as palm oil mill effluent (POME). In the process of palm oil milling, POME is generated through sterilization of fresh oil palm fruit bunches, clarification of palm oil and effluent from hydro-cyclone operations [1]. POME is a viscous brown liquid with fine suspended solids at pH ranging between 4 and 5 [2]. In general appearance, palm oil mill effluent (POME) is a yellowish acidic wastewater with fairly high polluting properties, with average of 25,000 mg/l biochemical oxygen demand (BOD), 55,250 mg/l chemical oxygen demand (COD) and 19,610 mg/l suspended solid (SS). This highly polluting wastewater can cause several pollution problems. Anaerobic digestion is the most suitable method for the treatment of effluents containing high concentration of organic carbon such as POME [1]. Anaerobic digestion is defined as the engineered methanogenic anaerobic decomposition of organic matter. It involves different species of anaerobic microorganisms that degrade organic matter [3]. In the anaerobic process, the decomposition of organic and inorganic substrate is carried out in the absence of molecular oxygen. The biological conversion of the organic substrate occurs in the mixtures of primary settled and biological sludge under anaerobic condition followed by hydrolysis, acidogenesis and methanogenesis to convert the intermediate compounds into simpler end products as methane (CH_4) and carbon dioxide (CO_2) [4], [5], and [6]. Therefore, the anaerobic digestion process offers great potential for rapid disintegration of organic matter to produce biogas that can be used to generate electricity and save fossil energy [7]. The suggested anaerobic treatment processes for POME include anaerobic suspended growth processes, attached growth anaerobic processes (immobilized cell bioreactors, anaerobic fluidized bed reactors and anaerobic filters), anaerobic blanket processes (up-flow anaerobic sludge blanket reactors and anaerobic baffled reactors), membrane separation anaerobic treatment processes and hybrid anaerobic treatment processes.

Over the past 20 years, the technique available for the treatment of POME in Malaysia has been biological treatment, consisting of anaerobic, facultative and aerobic pond systems [8, 9]. Anaerobic digestion has been employed by most palm oil mills as their primary treatment of POME [10]. More than 85% of palm oil mills producers in Malaysia have adopted the ponding system for POME treatment [11] due to its low capital and operating costs, while the rest opted for open digesting tanks [12]. These methods are regarded as conventional POME treatment method whereby long retention times and large treatment areas are required. High-rate anaerobic bio-reactors have also been applied in laboratory-scaled POME treatment such as up-flow anaerobic sludge blanket (UASB) reactor [13]; up-flow anaerobic filtration [14]; fluidized bed reactor [15, 16] and up-flow anaerobic sludge fixed-film (UASFF) reactor [2]. Anaerobic contact digester [12] and continuous stirred tank reactor (CSTR) have also been studied for POME treatment [17]. Other than anaerobic digestion, POME has also been treated using membrane technology [14, 18, 19], [20] and [21]. (POME's) chemical oxygen demand (COD) and biochemical oxygen demand (BOD) are very high; COD values greater than 80,000 mg/l and; pH values in the acidic range between (3.8 and 4.5) are frequently reported and the incomplete extraction of palm oil from the palm nut can increase COD values substantially. The effluent is non-toxic because no chemicals are added during the oil extraction process [22, 23, and 24]. (POME) is a brownish colloidal suspension, characterised by high organic content, and high temperature (70-80 °C) [25]. Most commonly, palm oil mills have already suggested use of anaerobic digesters for the primary treatment [26, 27]. The three widely used kinetic models considered in this study are shown in Table 1. The traditional ways for wastewater treatment from both economic (high cost) and environmental (harmful) disadvantages, this paper aims to introduce a new design technique of ultrasonic-membrane anaerobic system (UMAS) in treating POME and producing methane and to determine the kinetic parameters of the process, based on three known models; Monod [28], Contois [29] and Chen and Hashimoto[30].

Kinetic Model	Equation 1	Equation 2
Monod	$U = \dfrac{k\,S}{k_s + S}$	$\dfrac{1}{U} = \dfrac{K_s}{K}\left(\dfrac{1}{S}\right) + \dfrac{1}{k}$ [28]
Contois	$U = \dfrac{U_{max} \times S}{Y(B \times X + S)}$	$\dfrac{1}{U} = \dfrac{a \times X}{\mu_{max} \times S} + \dfrac{Y(1+a)}{\mu_{max}}$ [29]
Chen & Hashimoto	$U = \dfrac{\mu_{max} \times S}{Y\,K\,S_o + (1-K)\,S\,Y}$	$\dfrac{1}{U} = \dfrac{Y\,K\,S_o}{\mu_{max}\,S} + \dfrac{Y(1-K)}{\mu_{max}}$ [30]

Table 1. Mathematical expressions of specifics substrate utilization rates for known kinetic models

2. Materials and methods

Raw POME was treated by UMAS in a laboratory digester with an effective 200-litre volume. Figs. 1&2 presents a schematic representation of the ultrasonicated-membrane anaerobic system (UMAS) which consists of a cross flow ultra-filtration membrane (CUF) apparatus, a centrifugal pump, and an anaerobic reactor. 25 KHz multi frequency ultrasonic transducers (to create high mechanical energy around the membrane to suspends the particles) connected into the MAS system. The ultrasonic frequency is 25 KHz, with 6 units of permanent transducers and bonded to the two (2) sided of the tank chamber and connected to one (1) unit of 250 watts 25 KHz Crest's Genesis Generator. The UF membrane module had a molecular weight cut-off (MWCO) of 200,000, a tube diameter of 1.25 cm and an average pore size of 0.1 µm. The length of each tube was 30 cm. The total effective area of the four membranes was 0.048 m². The maximum operating pressure on the membrane was 55 bars at 70 ºC, and the pH ranged from 2 to 12. The reactor was composed of a heavy duty reactor with an inner diameter of 25 cm and a total height of 250 cm. The operating pressure in this study was maintained between 2 and 4 bars by manipulating the gate valve at the retentate line after the CUF unit.

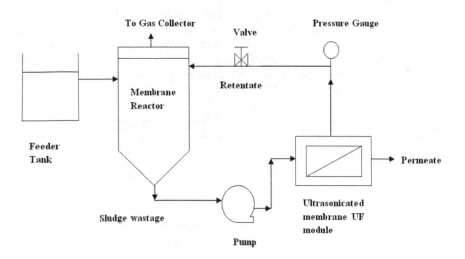

Figure 1. Experimental set-up

Figure 2. Schematic for Ultrasonicated membrane anaerobic system (UMAS)

2.1. Palm oil mill effluent

Raw POME samples were collected from a palm oil mill in Kuantan-Malaysia. The wastewater was stored in a cold room at 4°C prior to use. Samples analysed for chemical oxygen demand (COD), total suspended solids (TSS), pH, volatile suspended solids (VSS), substrate utilisation rate (SUR), and specific substrate utilisation rate (SSUR).

2.2. Bioreactor operation

The ultrasonicated membrane anaerobic system, UMAS Performance was evaluated under six different inflow conditions (six steady-states) with influent COD concentrations ranging from (67,000 to 91,400 mg/l) and organic loading rates (OLR) between (0.5 and 9.5 kg COD/m³/d). In this study, the system was considered to have achieved steady state when the operating and control parameters were within ± 10% of the average value. A 20-litre water displacement bottle was used to measure the daily gas volume. The produced biogas contained only CO_2 and CH_4, in order to collect pure CH_4, the addition of sodium hydroxide solution (NaOH) will absorb CO_2 effectively and the remaining will be methane gas (CH_4).

Steady State (SS)	1	2	3	4	5	6
COD feed, mg/L	67000	79000	82400	86000	90000	91400
COD permeate, mg/L	980	1940	1650	1980	2200	3000
Gas production (L/d)	280.5	357	377	395	470	540
Total gas yield, L/g COD/d	0.29	0.38	0.65	0.77	0.82	0.88
% Methane	79	75.5	70.2	71.8	70.6	68.5
Ch_4 yield, l/g COD/d	0.29	0.32	0.50	0.54	0.56	0.59
MLSS, mg/L	12960	13880	15879	17700	20000	25600
MLVSS, mg/L	10091	10950	12624	14638	17000	22528
% VSS	77.86	78.89	79.50	82.70	85.00	88.00
HRT, d	480.3	76.40	20.3	8.78	7.36	5.40
SRT, d	860	320	132	32.6	14.56	10.6
OLR, kg COD/m³/d	0.5	1.5	3	5.5	8.5	9.5
SSUR, kg COD/kg VSS/d	0.185	0.262	0.266	0.274	0.315	0.321
SUR, kg COD/m3/d	0.0346	0.8454	3.3028	5.6657	7.7753	9.4528
Percent COD removal (MAS)	**96.5**	**96.0**	**95.8**	**95.4**	**94.9**	**94.8**
Percent COD removal (UMAS)	**98.5**	**97.5**	**98.0**	**97.7**	**97.6**	**96.7**

Table 2. Summary of results (SS: steady state)

Model	Equation	R^2 (%)
Monod	$U^{-1} = 2025\,S^{-1} + 3.61$ $K_s = 498$ $K = 0.350$ $\mu_{Max} = 0.284$	98.9
Contois	$U^{-1} = 0.306\,X\,S^{-1} + 2.78$ $B = 0.111$ $u_{Max} = 0.344$ $a = 0.115$ $\mu_{Max} = 0.377$ $K = 0.519$	97.8
Chen & Hashimoto	$U^{-1} = 0.0190\,S_o\,S^{-1} + 3.77$ $K = 0.006$ $a = 0.006$ $\mu_{Max} = 0.291$ $K = 0.374$	98.7

Table 3. Results of the application of three known substrate utilisation models

3. Results and discussion

3.1. Semi-continuous Ultrasonic-Membrane Anaerobic System (UMAS) performance

Table 2 summarises UMAS performance of six inflow rates all (at six steady-states), which were established at different HRTs and influent COD concentrations. The kinetic coefficients of the selected models were derived from Eq. (2) in Table 1 by using a linear relationship; the coefficients are summarised in Table 3. At steady-state conditions with influent COD concentrations of 67,000-91,400 mg/l, UMAS performed well and the pH in the reactor remained within the optimal working range for anaerobic digesters (6.7-7.8). At the first steady-state, the MLSS concentration was about 12,960 mg/l whereas the MLVSS concentration was 10,091 mg/l, equivalent to 77.9% of the MLSS. This low result can be attributed to the high suspended solids contents in the POME. At the sixth steady-state, however, the volatile suspended solids (VSS) fraction in the reactor increased to 88% of the MLSS. This indicates that the long SRT of UMAS facilitated the decomposition of the suspended solids and their subsequent conversion to methane (CH_4); this conclusion supported by [31] and [32]. The highest influent COD was recorded at the sixth steady-state (91,400 mg/l) and corresponded to an OLR of 9.5 kg COD/m^3/d. At this OLR the, UMAS achieved 96.7% COD removal and an effluent COD of 3000 mg/l. This value is better than those reported in other studies on anaerobic POME digestion [33, 34]. The three kinetic models demonstrated a good relationship (R^2 > 99%) for the membrane anaerobic system treating POME, as shown in Figs. 2-5. The Contois and Chen & Hashimoto models performed better, implying that digester performance should consider organic loading rates. These two models suggested that the predicted permeate COD concentration (S) is a function of influent COD concentration (S_o). In Monod model, however, S is independent of S_o. The excellent fit of these three models (R^2 > 97.8%) in this study suggests that the UMAS process is capable of handling sustained organic loads between 0.5 and 9.5 kg m^3/d.

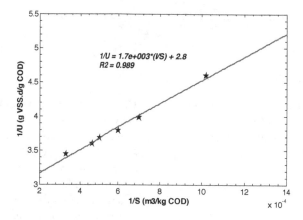

Figure 3. The monod model.

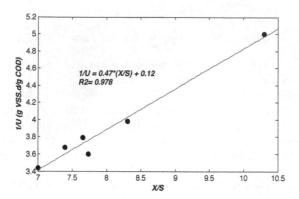

Figure 4. The Contois model.

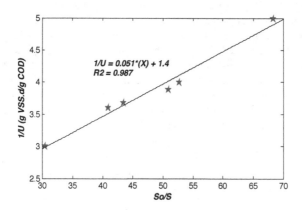

Figure 5. The Chen and Hashimoto mdel.

Fig.6 shows the percentages of COD removed by UMAS at various HRTs. COD removal efficiency increased as HRT increased from 5.40 to 480.3 days and was in the range of 96.7 % - 98.5 %. This result was higher than the 85 % COD removal observed for POME treatment using anaerobic fluidised bed reactors [35] and the 91.7-94.2 % removal observed for POME treatment using MAS [36]. The COD removal efficiency did not differ significantly between HRTs of 480.3 days (98.5%) and 20.3 days (98.0%). On the other hand, the COD removal efficiency was reduced shorter HRTs; at HRT of 5.40 days, COD was reduced to 96.7 %. As shown in Table 2, this was largely a result of the washout phase of the reactor because the biomass concentration increased in the system.

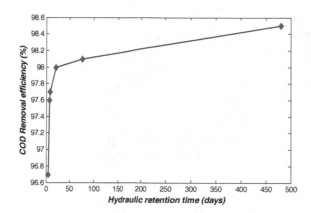

Figure 6. COD removal efficiency of UMAS under steady-state conditions with various hydraulic retention times.

3.2. Determination of bio-kinetic coefficients

Experimental data for the six steady-state conditions in Table 2 were analysed; kinetic coefficients were evaluated and are summarised in Table 3. Substrate utilisation rates (SUR); and specific substrate utilisation rates (SSUR) were plotted against OLRs and HRTs. Fig. 7 shows the SSUR values for COD at steady-state conditions HRTs between 5.40 and 480.3 days. SSURs for COD generally increased proportionally HRT declined, which indicated that the bacterial population in the UMAS multiplied [37]. The bio-kinetic coefficients of growth yield (Y) and specific micro-organic decay rate, (b); and the K values were calculated from the slope and intercept as shown in Figs. 8 and 9. Maximum specific biomass growth rates (μ_{max}) were in the range between 0.248 and 0.474 d^{-1}. All of the kinetic coefficients that were calculated from the three models are summarised in Table 3. The small values of μ_{max} are suggestive of relatively high amounts of biomass in the UMAS [38]. According to [39], the values of parameters μ_{max} and K are highly dependent on both the organism and the substrate employed. If a given species of organism is grown on several substrates under fixed environmental conditions, the observed values of μ_{max} and K will depend on the substrates.

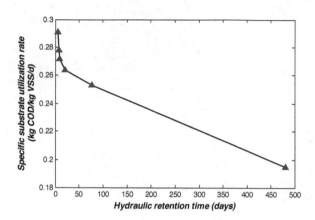

Figure 7. Specific substrate utilization rate for COD under steady-state conditions with various hydraulic retention times.

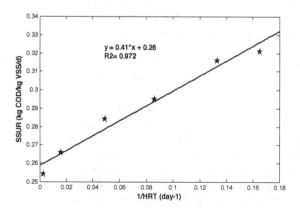

Figure 8. Determination of the growth yield, Y and the specific biomass decay rate, b

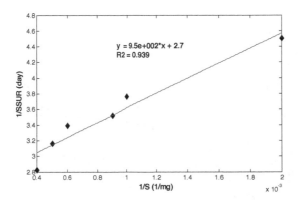

Figure 9. Determination of the maximum specific substrate utilization and the saturation constant, K

4. Gas production and composition

Many factors must be adequately controlled to ensure the performance of anaerobic digesters and prevent failure. For POME treatment, these factors include pH, mixing, operating temperature, nutrient availability and organic loading rates into the digester. In this study, the microbial community in the anaerobic digester was sensitive to pH changes. Therefore, the pH was maintained in an optimum range (6.8-7) (by addition of NaOH) to minimize the effects on methanogens that might biogas production. Because methanogenesis is also strongly affected by pH, methanogenic activity will decrease when the pH in the digester deviates from the optimum value. Mixing provides good contact between microbes and substrates, reduces the resistance to mass transfer, minimizes the build-up of inhibitory intermediates and stabilizes environmental conditions. This study adopted the mechanical mixing and biogas recirculation. Fig. 10 shows the gas production rate and the methane content of the biogas. The methane content generally declined with increasing OLRs. Methane gas contents ranged from 68.5% to 79% and the methane yield ranged from 0.29 to 0.59 CH_4/g COD/d. Biogas production increased with increasing OLRs from 0.29 l/g COD/d at 0.5 kg COD/m³/d to 0.88 l/g COD/d at 9.5 kg COD/m³/d. The decline in methane gas content may be attributed to the higher OLR, which favours the growth of acid forming bacteria over methanogenic bacteria. In this scenario, the higher rate of carbon dioxide; (CO_2) formation reduces the methane content of the biogas.

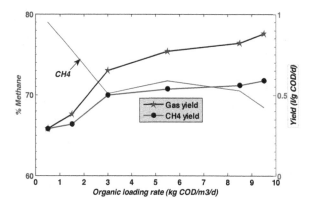

Figure 10. Gas production and methane content

5. Conclusions

The ultrasonic membrane anaerobic system, UMAS seemed to be adequate for the biological treatment of undiluted POME, since reactor volumes are needed which are considerably smaller than the volumes required by the conventional digester. UMAS were found to be an improvement and a successful biological treatment system that achieved high COD removal efficiency in a short period of time (no membrane fouling by introduction of ultrasonic). The overall substrate removal efficiency was very high-about 98.5%. The gas production, as well as the methane concentration in the gas was satisfactory and, therefore, could be considered (the produced methane gas) as an additional energy source for the use in the palm oil mill. Preliminary data on anaerobic digestion at 30 °C in UMAS showed that the proposed technology has good potential to substantially reduce the pollution load of POME wastewater. UMAS was efficient in retaining the biomass.The UMAS process will recover a significant quantity of energy (methane 79%) that could be used to heat or produce hot water at the POME plant.

Appendix A. nomenclature

COD: chemical oxygen demand (mg/l)

OLR: organic loading rate (kg/m^3/d)

CUF: cross flow ultra-filtration membrane

SS: steady state

SUR: substrate utilization rate (kg/m^3/d)

TSS: total suspended solid (mg/l)

MLSS: mixed liquid suspended solid (mg/l)

HRT: hydraulic retention time (day)

SRT: solids retention time (day)

SSUR: Specific substrate utilization rate (kg COD/kg VSS/d)

MAS: Membrane An aerobic System

UMAS: Ultrasonicated Membrane Anaerobic System

MLVSS: mixed liquid volatile suspended Solid (mg/l)

VSS: volatile suspended solids (mg/l)

MWCO: molecular weight Cut-Off

BLR: biological loading rate

U = specific substrate utilisation rate (SSUR) (g COD/G VSS/d)

S = effluent substrate concentration (mg/l)

S$_o$ = influent substrate concentration (mg/l)

X = micro-organism concentration (mg/l)

: Maximum specific growth rate (day^{-1})

K: Maximum substrate utilisation rate (COD/g/VSS.day)

: Half velocity coefficient (mg COD/l)

X: Micro-organism concentration (mg/l)

b = specific microorganism decay rate (day^{-1})

Y = growth yield coefficient (gm VSS/gm COD)

T: time

Author details

N.H. Abdurahman[1*], N.H. Azhari[2] and Y.M. Rosli[1]

*Address all correspondence to: nour2000_99@yahoo.com

1 Faculty of Chemical and Natural Resources Engineering, University of Malaysia Pahang-UMP, Malaysia

2 Faculty of Industrial Sciences and Technology, University of Malaysia Pahang-UMP, Malaysia

References

[1] Borja, R., C.J. Banks., E. Sanchez. Anaerobic treatment of palm oil mill effluent in a two-stage up-flow anaerobic sludge blanket (UASB) reactor. Journal of Biotechnology (1996a): 45, 125-135.

[2] Najafpour, G.D.., A.A.L, Zinatizadeh., A.R. Mohamed., M.Hasnain Isa., H.Nasrollah-zadeh. High-Rate anaerobic digestion of palm oil mill effluent in an upflow anaerobic sludge-fixed Film bioreactor. Biochemistry (2006): 41, 370-379.

[3] Cote, C., I.M. Daniel., S.Quessy. (2006). Reduction of indicator and pathogenic microorganisms by psychrophilic anaerobic digestion in swine slurries. Bioresource Technology 97, 686-691.

[4] Gee, P.T., N.S. Chua. (1994). Current status of palm oil mill effluent by watercourse discharge. In: PORIM National Palm Oil Milling and Refining Technology Conference.

[5] Guerrero, L., F. Omil., R.Mendez., J.M. Lema. (1999). Anaerobic hydrolysis and acidogenesis of wastewater from food industries with high content of organic solids and protein. Water Resource Journal 33, 3281-3290.

[6] Gerardi, M.H.. The microbiology of Anaerobic Digesters. Wiley-Interscience, New Jersey (2003): pp. 51-57.

[7] linke, B. (2006). Kinetic study of thermophilic anaerobic digestion of solid wastes from potato processing. Biomass and Bioenergy 30, 892-896.

[8] Chooi, C.F. Ponding system for palm oil mill effluent treatment. PORIM 9, (1984): 53-62.

[9] Ma. A.N. Treatment of palm oil mill effluent. Oil Palm and Environment: Malaysian Perspective. Malaysia Oil Palm Growers' Council. (1999): p.277.

[10] Tay, J.H. Complete reclamation of oil palm wastes. Resources Conservation and Recycling 5, (1991): 383-392.

[11] Ma. A.N.; S.C. Cheah., M.C. Chow. Current status of palm oil processing wastes management. In: Waste Management in Malaysia: Current status and Prospects for Bioremediation, (1993): pp. 111-136.

[12] Yacop, S., M.A. Hassan., Y.Shirai., M.Wakisaka., S.Subash. Baseline study of methane emission from open digesting tanks of palm oil mill effluent treatment. Chemosphere (2005): 59, 1575-1581.

[13] Borja, R.; C.J. Banks. Anaerobic digestion of palm oil mill effluent using an up-flow anaerobic Sludge blanket reactor. Biomass and Bioenergy (1994a): 6, 381-389.

[14] Borja, R.; C.J. Banks. Treatment of palm oil mill effluent by up-flow anaerobic filtration. Journal of Chemical Technology and Biotechnology (1994b): 61, 103-109.

[15] Borja, R.; C.J. Banks. Response of anaerobic fluidized bed reactor treating ice-cream wastewater to organic, hydraulic, temperature and pH shocks. Journal of Biotechnology (1995a): 39, 251-259.

[16] Borja, R.; C.J. Banks. Comparison of anaerobic filter and an anaerobic fluidized bed reactor treating palm oil mill effluent. Process Biochemistry (1995b): 30, 511-521.

[17] Ibrahim, A.; B.G. Yeoh., S.C. Cheah., A.N. Ma., S.Ahmad., T.Y. Chew., R.Raj., M.J.A. Wahid. Thermophilic anaerobic contact digestion of palm oil mill effluent. Water Science and Technology (1984): 17, 155-165.

[18] Chin, K.K. Anaerobic treatment kinetics of palm oil sludge. Water Research (1981): 15, 199-202.

[19] Ahmad, A.L.; M.F. Chong., S.Bhatia., S.Ismail. Drinking water reclamation from palm oil mill effluent (POME) using membrane technology. Desalination (2006): 191, 35-44.

[20] Ahmad, A.L.; M.F. Chong., S.Bhatia. Mathematical modeling of multiple solutes system for reverse osmosis process in palm oil mill effluent (POME) treatment. Chemical Engineering Journal (2007):132, 183-193.

[21] Fakhru'l-Razi. A. Ultrafiltration membrane separation for anaerobic wastewater treatment. Water.Sci. Technol (1994): 30 (12): 321-327.

[22] Abdurahman, N.H.; Y.M. Rosli., N.H. Azhari. Development of a membrane anaerobic system (MAS) for palm oil mill effluent (POME) treatment. Desalination (2011): 266, 208-212.

[23] Ma, A.N.; HA. Halim. Management of palm oil industrial wastes in Malaysia. Malaysia: Palm Oil Research Institute of Malaysia (PORIM)-Ministry of Primary Industries. (1988).

[24] Polprasert, C. Organic waste recycling. (1989). New York: John Wiley & Sons.

[25] Singh, . G.; LK, Huan; T, Leng; Kow DL. Oil palm and the environment. (1999).SDN. Bhd, Kuala Lumpur: Sp-nuda Printing.

[26] Anon. Biogas plants treating palm oil mill effluent in Malaysia. (1995). Rapa: Rural Energy.

[27] Tay, J. H. Complete reclamation of palm oil wastes. Resources conservation and Recycling 5, (1991): 383-392.

[28] Idris, A.B. and A. Al-Mamun. Effect of scale on the performance of anaerobic fluidized bed reactors (AFBR) treating palm oil mill effluent, Proc. Fourth International Symposum on Waste Management Problems in Agro-Industry, Istanbul, Turkey : (1998): 206-211.

[29] Monod, J. Growth of bacteria cultures. Annu Rev Microbial. (1949): 3:371-394.

[30] Contois, DE. Kinetics of bacteria growth: relationship between population density and space growth rate of continuous Cultures. J. Gen Microbiol. (1959): 21:40-50.

[31] Chen, Y. R.; A.G Hashimoto. Substrate Utilization Kinetic Model for Biological Treatment Processes, Biotechnol. Bioengn., (1980): 22, 2081-2095.

[32] Nagano, A.; E. Arikawa., and H. Kobayashi. (1992). The Treatment of liquor wastewater containing high-strength suspended solids by membrane bioreactor system. Wat. Sci Tech., (1992): 26 (3-4), 887-895.

[33] Borja-Padilla, R.; and C.J. Banks. Thermophilic semi-continuous anaerobic treatment of POME. (1993). Biotechnology Letters, 15(7), 761-766.

[34] Ng. W. J., K, K. Wong; and K. K. Chin. Two-phase anaerobic treatment kinetics of palm oil wastes. (1985).Wa. Sci. Tech., 19(5), 667-669.

[35] Idris, B.A.; and A. Al-Mamun. Effect of scale on the performance of anaerobic fluidized bed reactors (AFBR) treating palm Oil mill effluent, Proc. Fourth International Symposium on Waste Management Problems in Agro-Industry, (1998). Istanbul, Turkey: 206-211.

[36] Fakhru'l-Razi, A.; and M. J.M. M. Noor, Treatment of palm oil effluent (POME) with the membrane anaerobic system (MAS). (1999). *Wat. Sci. Tech.* 39(10-11): 159-163.

[37] Abdullah, A. G.; Liew, Idris. A., Ahmadun. F.R., Baharin, B. S., Emby, F., Noor, M. J. Megat., Mohd., Nour.A.H. A Kinetic study of a membrane anaerobic reactor (MAR) for treatment of sewage sludge. (2005). Desalination 183, 439-445.

[38] Zinatizadeh, A. A. L.; A. R. Mohamed; G. D. Najafpour. Kinetic evaluation of palm oil mill effluent digestion in a high rate up-flow anaerobic sludge fixed film bioreactor. (2006). Process Biochemistry Journal 41, 1038-1046.

[39] Grady, C. P.L.; H. C. Lim. Biological Wastewater Treatment: Theory and Applications. New York. (1980). Macel Dekker Inc.pp. 220-222, 870-876.

Permissions

The contributors of this book come from diverse backgrounds, making this book a truly international effort. This book will bring forth new frontiers with its revolutionizing research information and detailed analysis of the nascent developments around the world.

We would like to thank Dr. Nigel W.T. Quinn, for lending his expertise to make the book truly unique. He has played a crucial role in the development of this book. Without his invaluable contribution this book wouldn't have been possible. He has made vital efforts to compile up to date information on the varied aspects of this subject to make this book a valuable addition to the collection of many professionals and students.

This book was conceptualized with the vision of imparting up-to-date information and advanced data in this field. To ensure the same, a matchless editorial board was set up. Every individual on the board went through rigorous rounds of assessment to prove their worth. After which they invested a large part of their time researching and compiling the most relevant data for our readers. Conferences and sessions were held from time to time between the editorial board and the contributing authors to present the data in the most comprehensible form. The editorial team has worked tirelessly to provide valuable and valid information to help people across the globe.

Every chapter published in this book has been scrutinized by our experts. Their significance has been extensively debated. The topics covered herein carry significant findings which will fuel the growth of the discipline. They may even be implemented as practical applications or may be referred to as a beginning point for another development. Chapters in this book were first published by InTech; hereby published with permission under the Creative Commons Attribution License or equivalent.

The editorial board has been involved in producing this book since its inception. They have spent rigorous hours researching and exploring the diverse topics which have resulted in the successful publishing of this book. They have passed on their knowledge of decades through this book. To expedite this challenging task, the publisher supported the team at every step. A small team of assistant editors was also appointed to further simplify the editing procedure and attain best results for the readers.

Our editorial team has been hand-picked from every corner of the world. Their multi-ethnicity adds dynamic inputs to the discussions which result in innovative outcomes. These outcomes are then further discussed with the researchers and contributors who give their valuable feedback and opinion regarding the same. The feedback is then collaborated with the researches and they are edited in a comprehensive manner to aid the understanding of the subject.

Apart from the editorial board, the designing team has also invested a significant amount of their time in understanding the subject and creating the most relevant covers. They scrutinized every image to scout for the most suitable representation of the subject and create an appropriate cover for the book.

The publishing team has been involved in this book since its early stages. They were actively engaged in every process, be it collecting the data, connecting with the contributors or procuring relevant information. The team has been an ardent support to the editorial, designing and production team. Their endless efforts to recruit the best for this project, has resulted in the accomplishment of this book. They are a veteran in the field of academics and their pool of knowledge is as vast as their experience in printing. Their expertise and guidance has proved useful at every step. Their uncompromising quality standards have made this book an exceptional effort. Their encouragement from time to time has been an inspiration for everyone.

The publisher and the editorial board hope that this book will prove to be a valuable piece of knowledge for researchers, students, practitioners and scholars across the globe.

List of Contributors

J. E. Cortés Muñoz, C. G. Calderón Mólgora, A. Martín Domínguez, S. L. Gelover Santiago and G. E. Moeller Chávez
Mexican Institute of Water Technology, Mexico

E. E. Espino de la O and C. L. Hernández Martínez
National Water Commission, Mexico

Tin-Chun Chu and Matthew J. Rienzo
Department of Biological Sciences, Seton Hall University, South Orange, NJ, USA

N.H. Abdurahman, Y.M. Rosli and N.H. Azhari
LebuhrayaTun Razak, Gambang, Kuantan, Malaysia
Faculty of Chemical and Natural Resources Engineering, University Malaysia Pahang, Malaysia

Huynh Viet Khai and Mitsuyasu Yabe
Can Tho University, Vietnam Kyushu University, Japan

N.H. Abdurahman and Y.M. Rosli
Faculty of Chemical and Natural Resources Engineering, University of Malaysia Pahang-UMP, Malaysia

N.H. Azhari
Faculty of Industrial Sciences and Technology, University of Malaysia Pahang-UMP, Malaysia